BIOTECHNIQUES OF ECOLOGY

By

Ashok Kumar

Lecturer

Department of Zoology

Bundelkhand University

Campus Department

Jhansi

D P H

DISCOVERY PUBLISHING HOUSE

NEW DELHI-110002

ISBN: 978-81-8356-115-0

Biotechniques of Ecology

Published by:

DISCOVERY PUBLISHING HOUSE
4383/4B, Ansari Road, Darya Ganj
New Delhi-110 002 (India)
Phone: +91-11-23279245; 23253475; 43596065
Mobile: +91 9811179893 / +91 9871656464
E-mail: discoverybooksindia@gmail.com
orderdphbooks@gmail.com
namitwasan9@gmail.com
web: www.discoverypublishinggroup.com

Printed at:
Infinity Imaging Systems
Delhi

Preface

The present title "**Biotechniques of Ecology**" aims to provide a handbook of ecological techniques pertinent for the study of animals in their natural environment. There are, however, certain methods that are peculiar to the ecologist, those concerning the central themes of the subject, the measurement, description and analysis of samples. A great deal of careful observation is needed to identify the various organisms and to begin to note the major food chains. After the food chains have been identified, the ecologist can make more detailed measurements. Various environmental parameters like physico-chemical nature of three basic habitats—soil, water and air: temperature, light, humidity, pH, wind velocity etc., can be measured.

To acquire ecological knowlege about any kind of ecosystem of the biosphere, the ecologists have to use a variety of tools and techniques of other sciences like biology, chemistry, physics, mathematics, statistics, climatology, sociology, and economics. The present book deals with the various tools and techniques used in the ecological studies. The methods are dealt stepwise to make them easy and understandable.

There can be no claim to originality except in the manner of treatment and much of the information has been obtained from the books and scientific journals available in different libraries.

The author expresses his thanks to his friends and colleagues whose constant inspiration have initiated him to bring out this book.

The author expresses his gratitute to Mr. Wasan and staff of M/s Discovery Publishing House for their whole hearted co-operation in the publication of this book.

Author

CONTENTS

1

Hydrographic Mapping and Morphometry

MAPPING OF THE SITE

A map of the lake with appropriate scale may be produced from the competent authority. If it is not available it may be prepared by following procedure.

Traverse the area in several direction and put in stakes or flags at a number of points all around the lake. Make sure that one stake is visual from the other. Now measure the distance between stakes, 1-2, 2-3, 1-3, 2-1 etc. Tabulate as follows

Points	Distances (m)	Remarks

Measurements should be continued till there are least two measurements from each points and the end point coincides with the starting one.

Now take a graph sheet and make points (stakes) on it from the measured distances.

Pond Mapping by Plane Table Method

Select the longest relatively straight section of the lake shore for establishing a base line. Fix steel stakes near the shore for preparation of one end of base line. Most of the pond should be visible from this point. Repeat for other end of base line at a point at least 10 M from first stake and approximately the same distance from shore as the first. Make the base line as long as possible

Drive a wooden stake at water's edge at every major change in shoreline configuration. Maximum distance between any two wooden stakes should probably not exceed 5 M. Tie a cloth flag to top of every fifth stake to facilitate counting.

Attach graph paper to plane table. Set up tripod directly over one metal stake at end of base line and plumb table center to the stake. Judge shape of pond and length of base line and determine appropriate scale. Determine position on map that represents end of base line where table is set up. Have flagman set stadia rod vertically on steel rod at other end of base line. Place zero mark on alidade against pin, and sight through alidade down base line up with stadia rod. Draw base line on map along edge of alidade for the appropriate length as determined by scale.

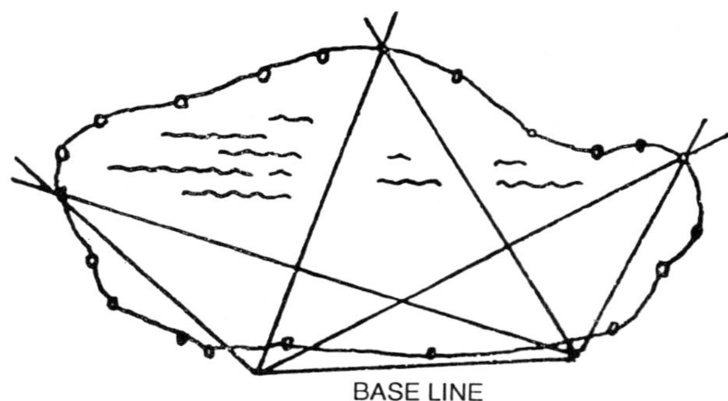

Figure 1.1: ***(a) Plane table map, the Base line, position and number of shoreline stakes. The shore line points are determined by triangulation this method is used for shallow lakes and ponds.***

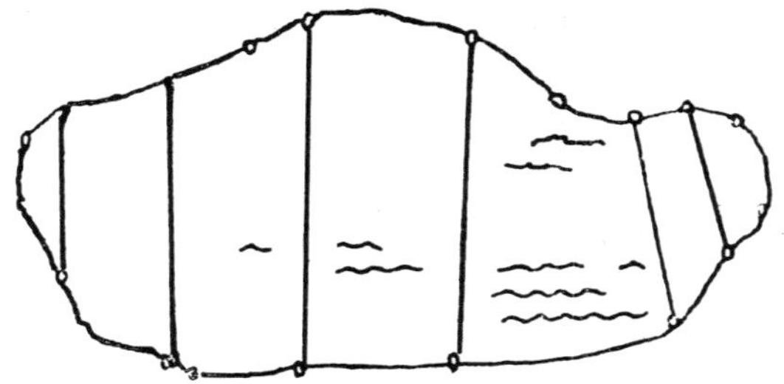

Figure 1.1: ***(b) Bottom contours in small ponds may be determined by soundings or along a calibrated rope stretched across.***

Now in a similar manner, progressively sight alidade around the pond on each wooden stake and draw line along edge of alidade. Number each line consecutively. The fifth line and every multiple of 5 should be sighted on a stake with flaf. Any deviation indicates a missed stake. After each flagged wooden stake has been sighted and the line drawn, resight on other end of base line.

The points where corresponding numerical lines, intersect is the position of each wooden stake marking the shoreline, connect these points with a smooth line. By doing this in the field, you are able to include minor changes in shoreline.

Pond Contour Map

Prepare the outline map of the water body as described above. Determine depth of the water at several points with the help of a boat and scaled sticks. For mapping deep ponds wherein sticks cannot be placed, the boat is rowed slowly in a straight line with the two land marks marked on the shore. A graduated string bearing a small hooked metal achor is thrown ahead in the pond from the moving boat and the depth is read on the string as soon as it stands vertical.

The *bethymetry* of a lake basin is most accurately determined from a continuous record of the basin contours with an accurate sonardevice (Depth sounders or Fathometer).

Mannual sounding of depth are made along transect between shoreline points with weighted, calibrated lines or, in shallow waters with metered sounding poles.

The depth measurements along transects are then plotted precisely to scale within the map of the shoreline. Points of equal depth at convenient intervals (e.g. *I M, 2M,* or 5M) are then interpolated between the points of known depth in several directions. These points of equal depth are then connected by contour lines.

The map should be constructed as follows in next page:

1. Enter all numbers representing sounding in the normal reading position using simple, clean-out numerals centred on the sounding position.

2. Draw submerged contours at desired intervals. Construct such contours by drawing a continuous line connecting all points

having same depth, pass contour line, through all numbers having exactly the value of the closed contour, where the exact chosen value is absent pass contour line between those sounding which are less and greater than the chosen values. Estimating position of contour line between the two values.

3. All contour line should be prepared of uniform thickness and lighter than shore time.

4. On every hydrographic map when finished there should appear certain essential information and explanations. Following information must appear on the map

(*i*) General title, (*ii*) Serial number, if one of the series, (*iii*) Geographic location, (iv) Scale of scales, (v) Official name *of* the lake, (vi) Compass figure, (*vii*) Low water datum and its use, (*viii*) Date of field work and map construction,

(*ix*) Names *of* field party and draftsman, (x) Abbreviations used and their meaning, (xi) Key to symbols used, (*xii*) Units used in sounding (feet, meters, fathorms), (*xiii*) Value of submerged contours.

5. Since water surface elevation fluctuate, causing depths to have a different value in the same place at different times values should be referred to some known elevation. It is a standard practice to reduce soundings to low-water datum for that body *of* water. To reduce all soundings to low-water datum it is necessary to know the surface elevation at the time sounding are made, then the difference between the low water datum and surface elevation is subtracted from the soundings as made in the field and the value now reduced should be entered upon the map.

Mapping of Stream

Only narrow and shallow streams can be surved by this method.

Measure a 50 meter or more stretch of the stream to be mapped. Mark at each interval of one meter. Mark also a line, called base line, thereafter on the graph paper representing the points on a suitable scale. Measure the distance from the each point to the near boundary of the stream, by joining the points the map of the stream can be constructed.

Stream profile:

The profile of any section of a stream can be prepared by recording the depth across the stream at small intervals prepare a graph between the distance and depth to obtain the profile.

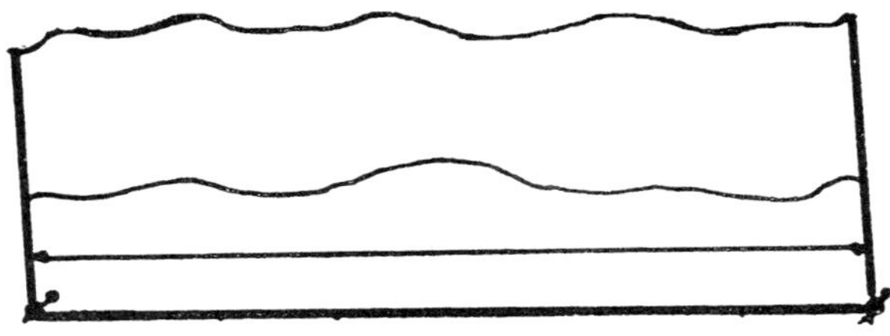

Figure 1.2 : ***(a) Mapping simple outline map of a stream.***

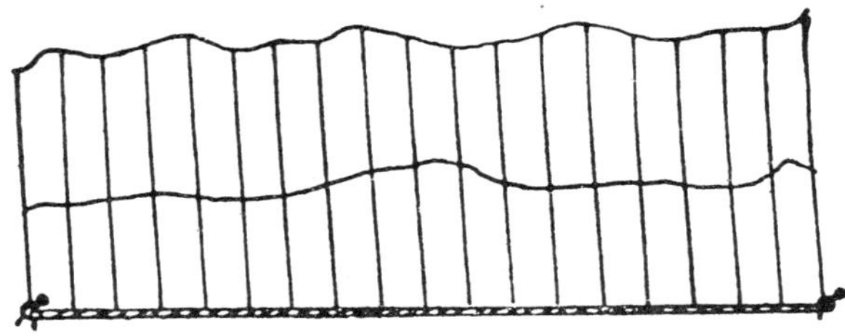

Figure 1.2 : ***(b) Mapping shoreline by measuring distance..***

MORPHOMETRY

Most limnological phenomena and productivity are directly related to the morphological features of the water basin. Therefore,

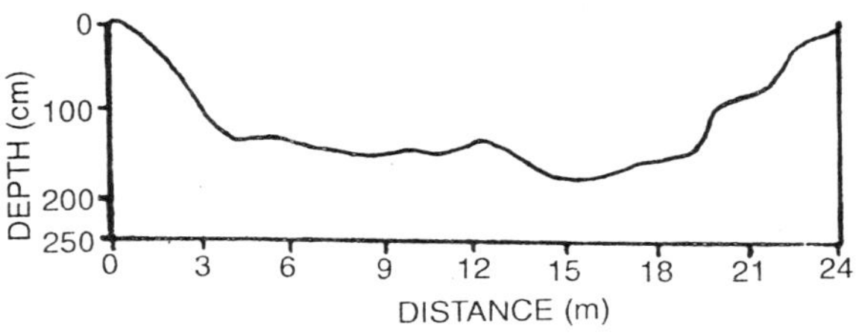

Figure 1.3 : ***Making depth profile in the cross section of a stream.***

certain morphometric features are interest to a limnologist beginning a study of any water. Morphometric measurements are based on good hydrographic maps and in general, the larger the map the more reliable the morphometric data that may be obtained from them.

Area by Polar Planimeter

Prepare map by placing it on hard, smooth surface. Tape it in place. Most maps will be too large to be covered by one cycle of the *planimeter*. These maps must be ruled off into halves or smaller segments and the areas of each summed for the total. Check the calibration of your instrument on the map paper being used. Use calibration device supplied by manufacturer, or carefully rule off a known area and trace with planimeter 3 times. Proceed with actual planimetery by tracing outline of lake or bottom contours onto a good grade of paper. From an area outside traced area, cut out a square of known area (a 9 inch square is usable) and weigh this piece of paper. Calculate weight per square inch. Cut out outline of entire lake and weigh. Calculate area by dividing weight of entire lake by weight of 1 square inch.

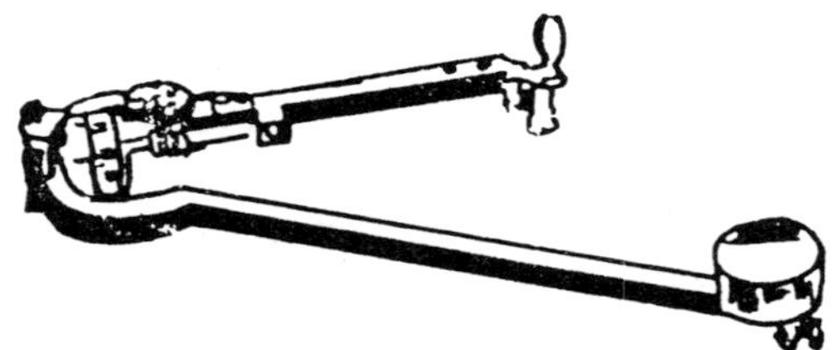

Figure 1.4 : ***(a)***

Volume by calculation

If a lake basin is considered as a cone, then the volume may be calculated by the appropriate equation (cone volume = 1.047 r 2h). However because the slopes of lake bottoms are rarely regular a better approximation for volume may be obtained by calculating and then summing the volumes of conical segments (frustra) with upper and lower surfaces delimited by the areas of sequential depth contours. The calculation is then as given by Welch (1958).

$$\text{Lake volume} = \sum (\text{frustrum volumes})$$

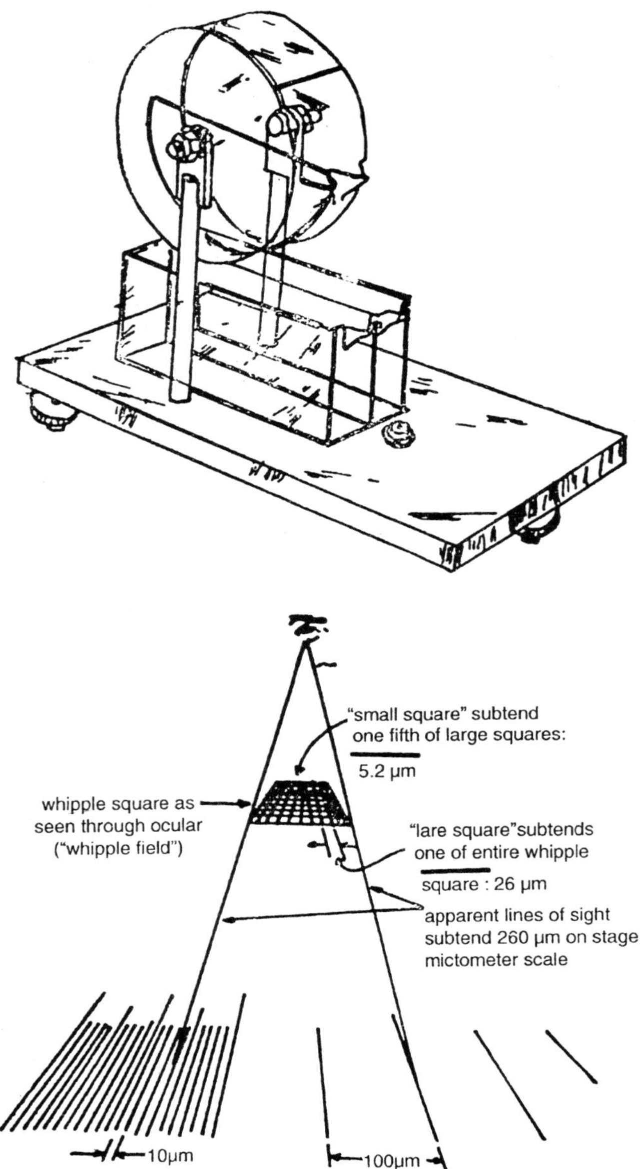

Figure 1.4 : ***B&C. Showing a simple type of polar planimeter.***

where

$$\text{frustrum volume} = \frac{h}{3}\left(a_1 + a_2 + \sqrt{a_1 a_2}\right)$$

h = depth of frustrum

a_1 = area of frustrum surface

a_2 = area of frustrum bottom.

Volume by Planimetery

Lake volume may also be determined by planimetric integration.

Using linear graph paper, plot the area at a given depth against that depth. Make the plot with the horizontal axis for area at the top and the vertical axis for depth at the left of the page. This places the 0-0 ordinate in the upper left corner.

Integrate the area beneath the curve by using a polar planimeter or by counting squares. Include those squares that are more than half within, and disregard those that are more than half outside the line. If equal dimensions for each square of the grid are used (that is each grid has dimensions of 1 m^2) in one direction versus volume. Summing the number of cubic meter volume grids under the curve will give the approximate total volume of the lake.

Shoreline length : cartometer method:

The map measure (cartometer is a convenient method for measuring) lengths of shorelines. It is also a delicate instrument and deserves care in handling :

1. Set dial by turning wheel to zero line. Draw a line of known length on the map paper and trace three times with the instrument to check its accuracy of calibration.

2. Set instrument to zero line, and carefully trace the shoreline of the lake. Watch carefully to see it the dial revolves more than one time. Record number to inches or centimeters, and convert to feet or meters per scale.

3. Repeat for each of the submerged contours.

Shoreline development :

Shoreline development is an index of the regularity of the

shoreline. For a lake that is a perfact circle, the shoreline development is 1. As the value departs from unity, irregularity is indicated. This value is calculated as follows

$$\text{Shoreline development (SLD)} = \frac{S}{2\sqrt{a\pi}}$$

where

S = length of shoreline

a = area of lake.

GENERAL MORPHOMETRIC MEASUREMENTS OF RESERVOIRS

For assessing and comparing the productivity of a reservoir it is required to collect certain morphometric data. In general reservoir being the artificial impondments, data on topography index mapping are capacity and geological information on the catchment area, available with the reservoir authorities. It should be produced before commencing the studies.

The details required are

(a) 1. Level (M.S.L.) Area (ha) and Volume (mc. ft.) for

(*i*) River bed (only level)

(*ii*) Dead storage

(*iii*) Full reservoir.

2. Catchment area

(*i*) Type

(*ii*) Area with average rainfall.

3. Rivers and streams following-Name and length.

Reservoir Levation

Collect the following information

Dam-(*i*) Type, (*ii*) Length, (*iii*) Height upto spill-way, (*iv*) Height upto ooest level, (*v*) No. and size of sluice gates, (*vi*) No. and size of river bed sludes.

Catchment area

1. Copy out an index map pertaining to reservoir which will

give catchment area in sq. km.

2. Geological data relating to catchment area may be had from competent authorities.

Submerged Countours of Reservoirs

Details of submerged us one generally incorporated in the project report pertaining to a particular reservoir would serve the purposes.

Derived Measures on Morphometry

Compute the following

(a) Length of the shore line it may be derived from reservoir map.

Method :

Pin an non-elastic thread along the reservoir boundary line on a large map and from the length of the thread, derive the length of the shore line making use of the scale of the map.

Water level and inflow-outflow data :

1. Mean monthly reservoir level with maximum and minimum (in meters).
2. Annual fluctuations of water level (in meters).
3. Monthly inflow into reservoir (f.cu.m.).
4. Monthly outflow from reservoir (f.cu.m.).

2

Estimation of Physical Factors

TEMPERATURE

Temperature is one of the most important factor and all life processes are accelerated or slowed down by temperature changes in the environment. It influences the solubility of gases and salts in water. The volume as well as density of the water depends upon temperature. The instruments which are used to measure temperature are mainly modifications of mercury expansion their moments. In recent days many electrical device like thermocouple and thermistors are used. Air temperature may be measured, by a normal mercury thermometer or a sixes type maximum and minimum thermometer.

Surface Temperature

Measurement of surface water temperatures is usually very important. Any good grade, simple type of mercury thermometer would serve the purpose. Any common chemical thermometer graduated to 0.2 degrees C can conveniently be used, but it should be secured well in a metal case to avoid loss in the field.

A good grade maximum and minimum thermometer will be very useful since it can be used for occasional recording of surface water temperature but also it can be adopted to the measurement of daily maximum and minimum temperature over a period of time. It may be mounted on a stationary object situated on the desired location and fixed enough below the surface of the water.

Sub-surface Temperature

Measurement of temperatures at various depths below the water surface require specially designed instruments for this purpose. Temperature often differ with depth and positions at which temperature is to be measured are remote from the observer. So it is necessary to understand the possible errors, such instruments when lowered meet rapidly increasing hydrostatic pressure and so many require correction for pressure as well as temperature.

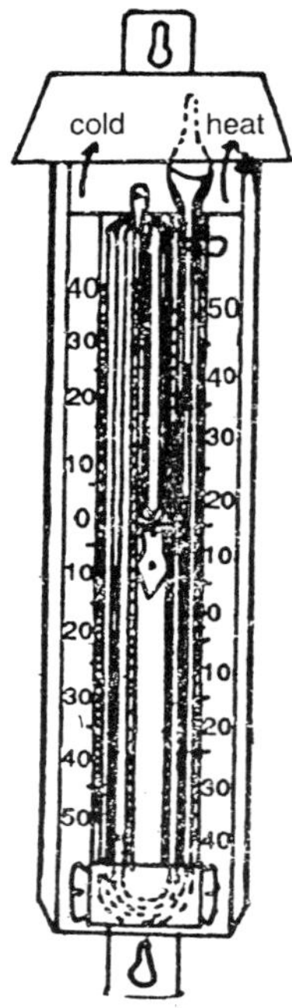

Figure 2.1 : *Maximum-minimum thermometer.*

Reversing Thermometers

This instrument consist of the main thermometer which registers the temperature and an ordinary thermometer which is used for correcting the change brought about by atmospheric temperature. The main thermometer has a *mercury reservoir* at one end of which is connected to the graduated capillary tube by a capillary loop with a construction. At the other end of the tube is an empty reservoir. For use, the thermometer frame is lowered into the water vertically to the desired depth. After few minutes a messenger is sent down the rope which reverses the thermometer frame. This reversing action breaks the mercury column at the point of construction and the mercury fills the empty bulb and a portion of the capillary tube. The level of mercury in the capillary indicates the temperature of water.

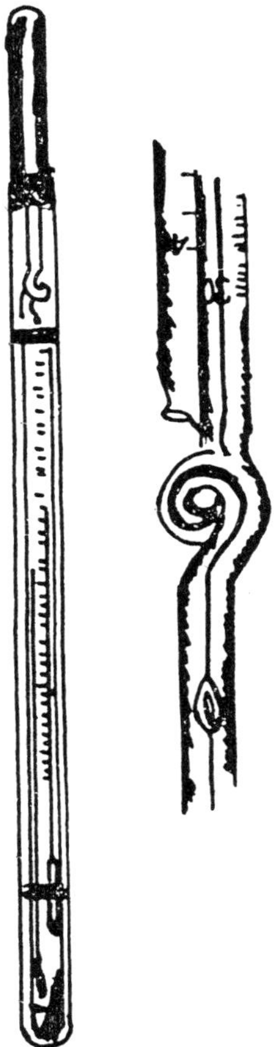

Figure 2.2 : ***Reversing thermometer.***

Thermistor :

This consist of a small capsule containing a complex *metal* oxide compound whose resistance decreases rapidly with increasing temperature. When used in a simple wheat stone bridge circuit it acts as a sensitive thermometer, and has the great advantage of being continuously recording. For each measurement, the valuable resistance R3 is adjusted until no reading is obtained on the galvanometer. When findings the temperature of water, the Thermistor (T) can be attached to a calibrated rod and low resistance leads fitted to it. It is important that these and all joints are well insulated with water proof material. The valuable resistance (R3) is provided by a rheostat fitted with a degree scale, while a small torch battery (B) and relatively inexpensive moving coil galvanometer (G) are the additional requirements.

Calibration :

A simple convention graph must be constructed in the laboratory by immersing the (Thermistor) in water of varying temperature (Measured by sensitive mercury thermometer) and plotting against corresponding values of R^3. It is important that while doing so same leads from the Thermistor showed by used as in the field.

TURBIDITY

In any water body, some materials are occurred in undissolved condition in the water, due to presence of these undissolved materials water becomes turbid. The *turbidity* of water may be more or less, it depends on dissolved materials which are present in the water. If these materials are more, the water will be more turbid.

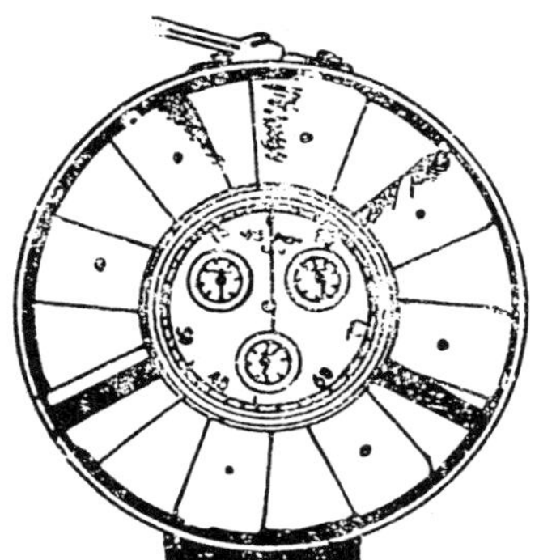

Figure 2.3 : ***Biram anemometer***

The turbidity of water is directly related to light penetration and visibility or transparency which can be measured by Secchi disc.

Measurement of Turbidity of Secchi Disc

The *Secchi disc* is a simple device used to turbidity and visibility of water. It consists of a weighted circular plate, 20 cms. in diameter, with the surface painted with opposing black and white quarters. It is attached to a calibrated line by a ring at the center so that when held by line, it hangs horizontally.

Procedure :

To determine the water turbidity, depth for Secchi disc visibility is determined. To determine the Secchi disc visibility depth, slowly lower the disc into the water u age of these two readings is taken for the final Secchi disc visibility depth.

Measurement of Turbidity by Turbidimeter

Turbidity of water is measured by *turbidimeter*. The standard unit for turbidity is the condition produced by 1ppm silica (Fuller's earth) in distilled water. Usually when turbidity *is* more than

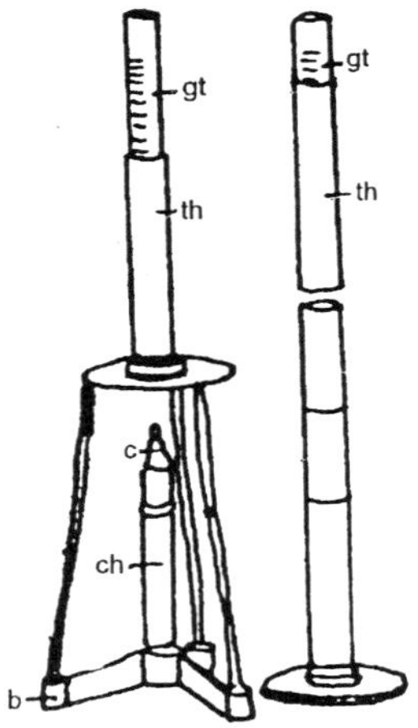

Figure 2.4 : ***Jackson Turbidimeter***

25 ppm a Jackson turbidity meter is used. It functions on the principal that the depth of a column of a flame of a standard candle disappear when viewed vertically through the column is a

measure of its turbidity. The candle should be lighted for few minutes at a time as the flame tends to increase in size as it burn, the candle resting on a tripod under the graduated tube. Installed in the candle support is a spring device which keeps the top of the candle at the required level (7.6 cm below glass tube) at all times so that the glass tube does not become hot. The graduations on glass tube directly read turbidity units ppm. For accurate measurements instrument should be operated in dark room. Higher turbidities can be measured by proper dilution.

COLOUR

Colour of aquatic body as seen by the naked eye is not the true representation of the hues inherent within the water itself. The measure used for one unit is the colour produced by 1.0 mg of platinum in 1 litre of distilled water. Standards are made by dissolving 1.245 gm of potassium chloroplatinate and 1 gm of crystallized cobaltous chloride in little distal water and adding 100 ml of conc. HC1 and final volume raised to 1 litre. This solution has colour value of 500 and lower standards (5, 10, 20, 25-70) can be prepared by dilution. If values are more than 70 the comparative Nessler jar method can be used to match and measure colours.

WATER CURRENT

The common method to measure water currents is by the *Ekman current meter*, which records both current rates and

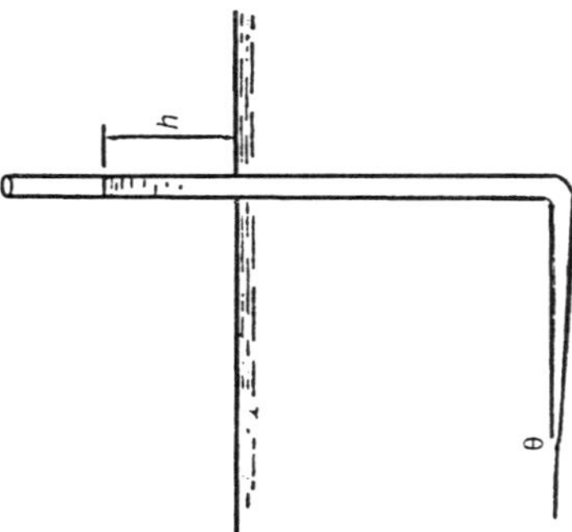

Figure 2.5 : ***Pitot tube in Simplest form.***

direction of current. Pitot Tube (Figure 2.5) is an 'L' shaped glass tube is also used and represents one of the simplest method to record water current velocity and direction.

WIND VELOCITY

The wind velocity estimated by well known cup anemometer which works on the rotation principal. It can also be recorded by the wind mill *Biran anemometer* and maxim's pressure. Plate anemometer.

TRANSPARENCY/LIGHT PENETRATION AND LIGHT INTENSITY

Ruttner's barrier layer photocells are used for light penetration and light intensity measurement in water. When they are illuminated, an electric current proportional to the strength of

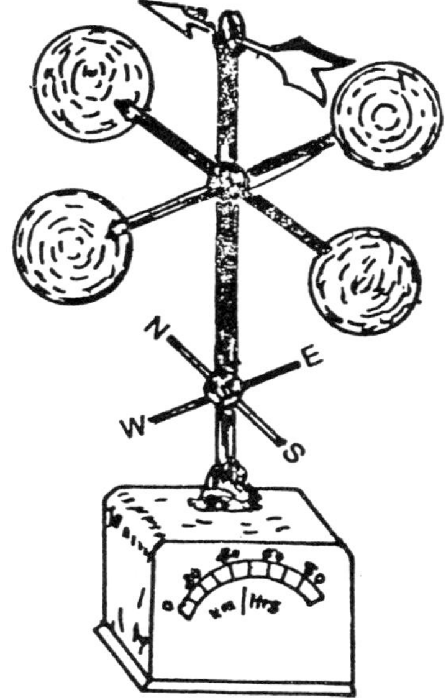

Figure 2.6 : ***A cup anemometer.***

illumination is produced, and these currents can be measured by galvanometer. The *photocell* is housed in a water proof box and is exposed inside the water at desired depths on a cable. The galvanometer is kept in the boat, the scale of the galvanometer

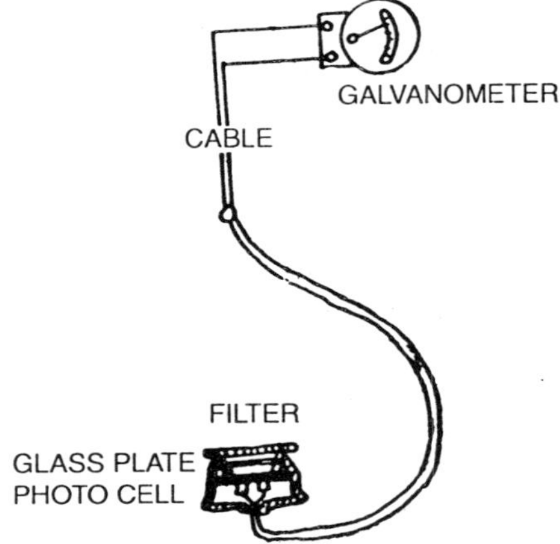

Figure 2.7 : *Barrier Layer Photoelectric Cell.*

being properly calibrated to show light intensity in lux. For crude recordings of Transparency the standard Sechhi disc may also the used.

HUMIDITY

For measuring humidity a *psychrometer* or a *hydrometer* is used.

The psychrometer consists of two measury thermometers. The wet bulb is covered by a muslin cloth which is exposed and connected to water source by a wick with the evaporation of water the temperature decreases. The humidity is inversely proportional to the drop in the thermometer. Equivalent humidity value can be recorded by reading the value from the humidity monogram.

Hygrometer

These are based on two principle of changes in the length of human hair due to humidity fluctuations. Humidity fluctuations from 0 to 100% cause a change of 2% change in the human hair. This can be recorded by using a *hygrothermograph*.

SPECIFIC CONDUCTIVITY

Specific conductance is a measure of a water's capacity to conduct an electric current. It is the reciprocal of resistance for which the standard unit is an ohm. Since conductance is the inverse of resistance, the unit of conductance is the *mho,* or in low-conductivity natural waters, the micro *mho*. Because the measurement is made using two electrodes placed 1 cm apart, specific conductance is generally reported as micromhos per centimeter. The relationship of specific conductance to ionized matter concentration varies with both the quality and quantity of the ions present. However, at low concentrations the ions move and behave independently, and the conductance concentration relationship is almost linear. Temperature of the solution affects the ionic velocity, and thus, the specific conductance.

TABLE 2.1 : CONVERSION OF RECORDED CONDUCTIVITY VALUES TO SPECIFIC CONDUCTANCE

°C	Factor	°C	Factor	°C	Factor	°C	Factor
3	1.62	15	1.21	27			
4	1.58	16	1.19	28			
5	1.54	17	1 16	29			
6	1.50	18	1.14	30			
7	1.46	19	1.12	31			
8	1.42	20	1.10	32			
9	1.39	21	1.08				
10	1.36	22	1.06				
11	1.33	23	1.04				
12	1.30	24	1.02				
13	1.27	25	1.00				
14	1.24	26	0.98				

Calculation-Conductivity at 25°C = *observed conductance × cell constant × temp. factor*

The greatest value of specific conductance to the limnologist is the estimation, by this very simple procedure, of the total concentration of dissolved ionic matter in the water, which in turn is related to water fertility. Specific conductance may be estimated from conductance data by multiplying by some empirical factor, which usually varies from 0.5 to 1.0. (Table 2.1) The measurements are often recorded by using a conductivity bridge or a conductivity meter.

Apparatus :

(*i*) Conductivity meter with dip cell

(*ii*) Water bath at 18°C or 25°C, if conductivity meter lacks temperature compensation

(*iii*) Thermometer

(*iv*)2 beakers of 200 to 250 ml capacity.

Procedure :

(*i*) Connect lugs of dip cell to instrument (the lugs may be connected to either post). A cell constant of 1 is best for most natural waters ; other cells or ranges are available for weak or strong conductors.

(*ii*) Fill 2 beakers with water sample to a depth sufficient to cover the vent holes in the dip cell when immersed. Place in water both, and wait for equilibrium if necessary.

(*iii*) Measure temperature of the second beaker, and set temperature compensator on instrument to that temperature.

(*iv*) Rinse dip cell thoroughly in the first beaker, and then transfer dip cell to second beaker. Turn on instrument, and while gently moving the dip cell, keeping the holes submerged all the time, take a reading of specific conductance.

(*v*) Turn off instrument, remove and disconnect dip cell, rinse with distilled water, and air dry.

3

Estimation of Chemical Factors

pH (HYDROGEN ION CONCENTRATION)

The pH of a solution is a measure of its hydrogen ion activity and is the logarithm of the reciprocal of the hydrogen ion concentration. Thus it is important to remember that a change of one pH unit represents a tenfold change in hydrogen ion concentration ; for example a pH of 6 has 10 times the hydrogen ions of pH 7, pH 5 has 100 times the hydrogen ions of pH 7.

The pH of most natural water falls in the range of 4 to 9 and much more often in the range of to 8. The following methods are used for measurement of hydrogen-ion-concentration:

1. Colour charts method.
2. Universal indicator solution method.
3. pH meter.

Colour Charts Method

In this method, sets of graded colours representing ranges of pH values are used. Such colours are usually printed on paper in some convenient form.

Procedure :

Take a pH indicator paper strip and dip it into given water sample and compare the colour of strip with given graded colours chart. This gives approximate pH values.

Universal Indicator Solution

In this method universal indicator solution is used. A few drops of this solution are added into the sample. This develops a colour in sample. The colour of sample is compared with given standard chart.

Indicator solution develops different types of colour at different pH in sample. This gives an approximates value of pH.

pH Meter

Apparatus:

(*i*) pH meter

(*ii*) Centigrade thermometer

(*iii*) Beakers

(*iv*) Blot paper.

Reagents:

(*i*) pH buffer solution in the range of water sample being analyzed.

(*ii*) Distilled water.

Procedure:

(*i*) Prepare a buffer solution with a pH near the expected pH of the water sample (Buffer may be prepared **in** time laboratory may be purchased in many forms). Place 25 ml buffer solution in a small beaker.

(*ii*) Open the vent of glass reservoir of the fiber junction electrode.

(*iii*) Rinse the electrode with distilled water, and blot with soft paper.

(*iv*) Immerse the electrodes in the buffer solution, measure temperature of buffer solution, and set compensator to that temperature.

(*v*) Set function switch to pH, and adjust meter needle to the pH of the buffer solution. The instrument is now standardized. Return switch to stand by.

(*vi*) Rinse electrodes in distilled water and blot dry.

(*vii*) Immerse electrodes in sample, measure temperature of sample, and set compensator to this temperature.

(*viii*) Set function switch to pH, and very gently swirl (of use magnetic stirrer at very slow speed) until pH meter needle stops drifting. Read pH from meter scale. Return a switch to standby. Raise electrodes, rinse with distilled water and dry. The instrument is now ready for the most sample or for storage.

(*ix*) Storage of instrument. Close the glass reservoir to minimize electrolyte loss, and leave electrodes immersed in distilled water.

ESTIMATION OF DISSOLVED OXYGEN

The dissolved oxygen (DO) content of waters results from (1) the photosynthetic and respiratory activities of the biota in the open water, the benthos, and the aufwuchs, and (2) the diffusion gradient at the air water interface and distribution by wind driven mixing.

Winkler Method

Principle of determination:

This method depends on the formations of a loose pricipitate (floc) of manganous hydroxide. The oxygen dissolved in the water is rapidly absorbed by the manganous hydroxide, forming a higher oxide, which may be in the following form.

$$MnSO_4 + 2KOH = Mn(OH)_2 + K_2SO_4$$

$$2Mn(OH)_2 + O_2 = 2MnO(OH)_2$$

As it settles, the $Mn(OH)_2$ floc acts as a "gathering" agent for the DO. Upon acidification in the presence of iodide, iodine is released in a quantity equivalent to the DO present.

$$MnO(OH)_2 + 2KI + H_2O = Mn(OH)_2 + J_2 + 2KOH$$

The liberated iodine is then titrated with a standard sodium thiosulphate solution, with starch used as the indicator.

$$I_2 + 2S_2O_3^{-} = S_4O_6^{-} + 2I^{-}$$

Apparatus:

(*i*) Narrow mouth darkened bottles of good quality and well fitted ground glass stoppers.

(*ii*) Pipette 1 ml, 10 ml and 25 ml.

(*iii*) Burette

(*iv*) Conical flasks etc.

Reagents:

(*i*) *Manganese sulphate solution.*

Dissolve 120g $MnSO_4.4H_2O$ or 100g $MnSO_4.2H_2O$ or 91g $MnSO_4.H_2O$ indistilled water, filter, and dilute to 250 ml.

(*ii*) *Alkaline-iodide.*

Dissolve 125 g sodium hydroxide (NaOH) and 33.75 g sodium iodide (Nal) in distilled water and dilute to 250 ml.

Or

Dissolve 175g KOH and 37.5g kl in distilled water and dilute to 250 ml.

(*iii*) *Starch solution.*

Prepare an emulsion of 6g starch in a beaker with a small amount of distilled water. Pour this emulsion in 1 litre of distilled water, boil for a few minutes, and let settle overnight. Save the clear supernate, and preserve by adding 1.25g salicylic acid.

(*iv*) *Sulphuric acid*

Concentrate sulphuric acid is used.

(*v*) *Sodium thiosulphate stock solution* (1.ON).

Dissolve 248.2 g $Na_2S_2O_3.5H_2O$ in boiled and cooled distilled water and dilute to 1000 ml. Preserve by adding 5 ml chloroform or 1 g NaOH per liter.

(*vi*) *Standard sodium thiosulphate solution* (0.025 N).

Prepare by (a) dilute 25 ml sodium thiosulphate stock solution (I.ON) to 1000 ml. or (b) dissolved 6.205 g $N_2S_2O_3.5H_2O$ in freshly boiled and cooled distilled water and dilute to 1 litre. Standard sodium thiosulphate solution may be preserved by adding 5 ml chloroform or 0.4g NaOH per litre.

(*vii*) Standardization of sodium thiosulphate solution with bi-iodate.

Dissolve approximately 2g potassium iodide, free from iodate, in an flask with 100 to 150 ml distilled water, add 10 ml 1 + $9H_2SO_4$ followed by exactly 20 ml standard 0.025 N bi-iodate solution. Dilute to 200 ml and titrate the liberated iodine with the thiosulphate solution, adding starch towards the end of the titration, when a pale colour is reached. Exactly 20 ml 0.025 N thiosulphate should be required when the solutions under comparison are of equal strength. It is convenient to adjust the solution to exactly 0.025 N. One ml 0.025 N thiosulphate is equivalent to 0.2 mg oxygen.

Procedure:

(1) Fixation of dissolved oxygen

(2) Titration.

Fixation of Dissolved Oxygen

Fixation of dissolved oxygen must be performed in the field immediately after the sample has been secured. Any delay may result in error. Dissolved oxygen is fixed by following procedure.

(*i*) Take water sample in 250 ml or 300 ml volume sample bottle.

(*ii*) Remove stopper from sample bottle and add 1 cc of manganous sulphate solution by inserting tip of pipette just below water surface.

(*iii*) Add 1 cc alkaline-iodide solution in same manner as for $MnSO_4$.

(*iv*) Replace stopper and mix sample by investing bottle several times. Allow precipitate to settle for a few minutes. After settling of pricipitate, clear fluid should occupy upper portion of bottle.

(*v*) Using volumetric pipett, add 1 cc concentrated sulphuric acid by permitting it to run down neck of bottle, mix well by inverting bottle several times (at least 40 times). Allow sample to stand for at least 5-10 minutes.

Titration

Take 100 ml or 50 ml of sample (in which dissolved oxygen

has been fixed) in a flask and titrate with 0.025 N sodium thiosulphate solution to a very pale yellow colour, now add I ml starch solution (colour of sample turns blue) and titrate continue until the first disappearance of the blue colour. Record the used volume of sodium thiosulphate solution in ml.

Calculation:

$$\text{mg of } O_2 \text{ per litre} = \frac{\text{Used Vol. of titrant} \times 1000 \times .2}{\text{Vol. of sample}}$$

where 0.2 value represents, 1 ml of sodium thiosulphate equivalent to 0.2 mg of O_2.

ALKALINITY OF ESTIMATION

Free-carbondioxide

Carbondioxide is an end product of decomposing bacteria and of the respiratory process of plants and animals. It is also added to the water by the action of natural or pollution acids on bicarbonates.

Principle:

Free CO_2 reacts with sodium carbonate or sodium hydroxide to form sodium bicarbonate. Completion of the reaction can be indicated by the development of the pink colour characteristic of phenolphthalein indicator at the equivalence pH of 8.3.

Apparatus:

(i) Burette (ii) Conical flask etc.

Reagents:

(*i*) Phenolphthalein indicator solution.

Preparation of phenolphthalein indicator solution is same as in estimation of alkalinity.

(*ii*) Standard Sodium carbonate titrant (0.045 N)

Dissolve 0.02 g anhydrous, primary standard grade Na_2CO_3 that has been even dried at 140°C overnight in freshly boiled and cooled carbondioxide free water, and dilute to 250 ml store in a rubber stoppered Pyrex bottle (This solution not be kept more than

1 to 2 weeks).

Procedure:

(*i*) Take 100 ml sample in Nessler tube and add 10 drops of phenolphthalein indicator. If the colour of the sample turns to pink, free carbon dioxide absent.

(*ii*) If the sample remains colourless, titrate rapidly with standard sodium carbonate titrant, stirring gently with a glass rod until a faint pink colour appears permanently.

(*iii*) Record the use volume of titrant is M1.

Calculation:

$$\text{mg}CO_2 \text{ per litre} = \frac{\text{Used Vol. of titrant} \times 1000}{\text{ml of sample}}$$

Estimation of Carbonate and Bicarbonate

The alkalinity of water is its capacity to accept protons, stated another way, it is the quantity and kinds of compounds present that collectively shift the pH to the alkaline side of neutrality. Although the alkalinity of natural waters is generally the result of bicarbonates, it is usually expressed in terms of calcium carbonate. Three kinds of alkalinity are indicated hydroxide (OH), normal carbonate (CO_3^{-}), bicarbonate HCOB ($HCO_3^{=}$) ; the three are summed as total alkalinity. Carbonates and bicarbonates are common to most waters because carbonate minerals are abundant in nature and because contribution to alkalinity by hydroxides is rare in nature. The presence of hydroxides can usually be attributed to water treatment or to contamination. Expected total alkalinities in nature usually range from 45 to 200 mg/litre.

Principle:

There are five conditions of alkalinity possible in the. sample, carbonate, bicarbonate, or hydroxide alone, or a combination of carbonate and hydroxide or of carbonate and bicarbonate. Hydroxide and bicarbonate are not found together.

Alkalinity is determined by titrating the water sample with a standard solution of strong acid. The equivalency or end points of the titration are selected as the inflection points in the titration of sodium carbonate with H_2SO_4. These points will vary slightly with temperature, ionic concentration, and free carbondioxide concentration. This end point may be determined empirically by titration and is the pH where the derivative of pH/ml titrant is greatest. However, they are usually taken as pH 8.3 for the carbonate end point and pH 4.5 for the bicarbonate end point. The following reactions occur in titrations

$$CO_3^{-} + H^{+}HCO_3^{-} \text{(titration to pH 8.3)}$$

$$HCO_3^{=} \text{(from } CO_3^{=}) + H^{+}H_2O + CO_2$$

$$HCO_3^{-} \text{(natural)} + H^{+}H_2O + CO_2 \text{(titration to pH 4.5)}$$

Apparatus:

(*i*) Burette

(*ii*) Conical flasks etc.

Reagents :

(*i*) Phenolphthalein indicator solution:

Dissolve 0.5 g phenolphthalein in 50 ml of ethyl alcohol and add 50 ml of distilled water. Then add 0.02 N NaOH dropwise until a faint pink colour appears.

(*ii*) Standard Sulphuric acid solution (0.02 N):

Prepare stock solution, approximately 0.1 N by diluting 3 ml of concentrated H_2SO_4 to 1 litre. Dilute 200 ml of 0.1 N stock solution to 1,000 ml with carbondioxide free water to give 0.02 N solution. For accurate work this must be standardized against Na_2CO_3.

(*iii*) Methylorange indicator solution:

Dissolve 0.5 g methyl orange to litre of distilled water.

Procedure:

(a) *Phenolphthalein alkalinity*

(*i*) Take 50 or 100 ml of water sample and add 4 drops of phenolphthalein indicator solution. If a pink colour appears (indicates the presence of carbonate) titrate against standard sulphuric acid (.02 N) until colour disappears (pH 8.3) and record millilitres of acid used.

(b) Total alkalinity (by methyl orange indicator method):

(*ii*) If no pink colour appears after adding phenolphthalein or after phenolphthalein titration, add 2 drops of methyl orange indicator and titrate to a faint orange at pH 4.6 (and pink at below 4 pH). Record total used standard acid in ml. *Calculation*

(I) Phenolphthalein alkalinity

$$\text{as mg } CaCO_3 \text{ per litre} = \frac{\text{ml standard acid} \times 1000}{\text{ml of sample}}$$

(II) Total alkalinity

$$\text{as mg } CaCO_3 \text{ per litre} = \frac{\text{Total ml standard acid} \times 1000}{\text{ml of sample}}$$

BIOCHEMICAL OXYGEN DEMAND (BOD)

The rate of removal of oxygen by organisms using the organic matter, present in water is called biochemical oxygen demand. The parameter can easily be estimated although only an approximation but gives an index of organic pollution (or of sewage treatment plant efficiency). The *BOD test* is commonly made by measuring O_2 concentration in samples before and after incubation in dark at 20°C for 5 days. It is necessary to dilute and aerate the sample to ensure that not all the O_2 is used during the incubation. Sometimes a culture of bacteria is also added so that more of the organic matter is used up during incubation.

Estimation of Biochemical Oxygen Demand (BOD)

A simplified method for BOD estimation give some information about difference between lakes or rivers or about seasonal changes in the dissolved organic matter in one water body.

Apparatus:

(*i*) Narrow mouth, darkened glass bottles (125, 250, 300 ml).

(*ii*) Air incubator or water bath thermometrically controlled of 20 ± 1 C. All light should be excluded to prevent DO by algae in the sample.

(*iii*) Burette, Pipettes, etc.

Procedure:

If the O_2 content of one original sample is very low, the sample should be aerated for 5 or 10 minutes. Place portions of the sample in to 3 glass stoppered bottles, the remaining bottle in the dark at a standard temperatures (e.g. 20 C.) or at the temperature of the original sample for 1-5 days. Then determine the remaining oxygen. Substract this value from the original value of oxygen to give BOD

CHEMICAL OXYGEN DEMAND (COD)

Principle

The concentration of organic compounds in lake water can be estimated by their oxidability by oxidizing substances such as $K_2Cr_2O_7$.

Organic carbon and O_2 are indirectly related because the reaction is:

$$C_xH_{2y}O_2 + \left[x + \frac{(y-z)}{2}\right]O_2 \rightarrow X\,CO_2 + yH_2O.$$

so that

$$1.0\ M / 1\ Cr_2O_7^{-} \text{ is } 6.0\ N.$$

Ag_2SO_4 is used as a catalyst for the oxidation. It is particularly effective for short straight-chain alcohols and acids.

Reagents:

(*i*) Standard oxalic acid, 0.0125 N.

(*ii*) $K_2Cr_2O_7$, 1.000 N.

(*iii*) H_2SO_4 (s.g.=1.48, A.R.)

(*iv*) H_2SO_4 (1+1) A.R.

(*v*) Ferrous solution, 0.25 N.

(*vi*) NH_2SO_3H (Sulphuric acid), (dry, A.R.)

(*vii*) Diphenylamin indicator, 0.5%.

(*viii*) Ag_2SO_4 (dry, A.R.)

(*ix*) $HgSO_4$ (dry, A.R.)

Procedure:

Mix 70 ml of sample containing not more than 16 mg COD (as O_2) or an aliquot diluted to 70 ml, in a 250 ml flask with powdered $HgSO_4$ in the ratio $HgSO_4$_Cl^- =10.1 (Cl less than 2g/1). Add, with caution, 20 ml of H_ZSO_4 (*iiia*). Then add successively 100 mg of Ag_2SO_4 and 10.00 ml of $K_2Cr_2O_7$ Attach the condenser to the arlenmeyer flask, and put the flask in a water bath. Either boil for 3-6 hrs or autoclave the covered flasks at about 135°C (3 atmospheres pressure) for 3-6 hours. Coll the flask, and then :

Either—

Titrate the excess of $K_2Cr_2O_7$ with reagent using a dead stop end point titrimeter or a potentiometric titrimeter. Graph the meter readings against ml of reagent.

Or

Add 1 ml of indicator and titrate the excess of $K_2Cr_2O_7$ with reagent (iv). At the end point the colour changes sharply from turbid blue to brilliant green.

Run a blank using 70 ml of double-distilled H_2O, and all reagents.

Calculation:

$$\text{COD mg / l} = \frac{(a-b) \times N \times 8 \times 1000}{\text{ml of sample}}$$

where

COD = Chemical oxygen demand in mg/l O_2.

a = ml titrant used for blank (about 40 ml)

b = ml titrant used for sample

N = Normality of the titrant.

CHLORIDE

Chloride is one of the major anions in water and sewage. The salty taste in water is produced due to the chloride concentration which is highly valuable depends on the chemical composition of water.

A high chloride contents also exerts a effect on metallic pipes and structures as well's on agricultural plants.

Estimation of Chloride

Principle:

In a natural or slightly alkaline solution, potassium chromate can indicate the end point of the silver nitrate titration of chloride. Silver nitrate is quantitatively precipitated before silver chromate is formed.

Apparatus:

(*i*) Burette

(*ii*) Conical flask etc.

Reagents:

(*i*) Silver nitrate solutions (0.0141 N) Dissolve 2.39 g $AgNO_3$ in 1 litre distilled water.

(*ii*) Potassium chromate indicator.

Dissolve 50 g $K_2Cr_2O_7$ in distilled water. Add $AgNO_3$ to produce a slight red ppt. After it has stood at all overnight, filter and dilute to 1 litre with distilled water.

Procedure:

(*i*) Take 50 or 100 ml of water sample and add 4-5 drops of potassium chromate indicator solution.

(*ii*) Now titrate the sample against the .0141 N $AgNO_3$ solution till pinkish-yellow colour appears.

(*iii*) Record the vol. of used titrant in ml.

Calculation:

$$\text{Mg of chloride per litre} = \frac{\text{Used Vol. of titrant} \times a \times b \times c}{\text{Vol. of water sample}}$$

where

a = 35.466 (atomic wt. of chloride)

b = .0141 (N *of* $AgNO_3$)

c = 1000 ml.

HARDNESS

Water that contains salts of calcium and magnesium and, to a lesser extent, other polyvalent metals such as iron, aluminium and manganese, referred to as *hard water*. *Hardness* is now best defined as a characteristic of water representing the total concentration of calcium and magnesium ions expressed as milligrams of $CaCO_3$ per litre. When other ions are present in insignificant amounts, the hardness will be equal to or less than the sum of the carbonate and bicarbonate alkalinities and is termed carbonate hardness. If the hardness exceeds the sum of these alkalinities, the presence of other ions indicated, and the excess is expressed as non-carbonate hardness.

Estimation of Hardness

Principle:

The criochrome black T indicator used in this titration produces a red colour in the presence of calcium and magnesium ions. The disodium salt of EDTA forms a stable, colourless complex with these ions, effectively removing them from solution. Thus by titration of the sample containing the indicator with EDTA, the calcium any magnesium ions are quantitatively removed. When all ions are removed, the indicator changes to a bright blue colour.

Apparatus:

(*i*) Conicle flasks

(*ii*) Burette and Pipetees.

Reagents:

(*i*) Amonia buffer solution.

(*ii*) Eriochrome Black Tindicator

(*iii*) Standard EDTA titrant (8.01 M):

Dissolve 0.3723 g Na_2 EDTA-dihydrate in distilled water and dilute to 100 ml. (Check by titrating against a standard calcium solution : 1.00 ml =1.00 mg $CaCO_3$ =0.4008 mg Ca.

Procedure:

(*i*) Take 50 ml water sample and add 1 to 2 ml of ammonia buffer solution to bring pH to 10.0 or 10.1.

(*ii*) Add approximately 0.1 g indicator powder.

(*iii*) Titrate with standard EDTA solution, continuously till the colour turns red to violet or bright blue. This indicates the end point.

(*iv*) Record the volume of used titrant.

Calculation:

EDTA hardness as mg $CaCO_3$ per litre

$$= \frac{\text{Used Vol. of titrant} \times A \times B}{\text{Vol. of sample}}$$

where

A = mg of $CaCO_3$ equivalent to 1.0 ml EDTA titrant

= 1 mg. so A=1

B = 1000 ml.

CALCIUM

Calcium can be leached from practically all rocks but is much more prevalent in water from regions with deposits of limestone, dolomite and gypsum. Regions where granite or silicious send predominate have very low calcium levels in the waters. Waters with a concentration of 10 mg of less per litre are usually oligotrophic, while waters with 25 mg or mere per jitre are usually distinctly outrophic,

Estimation of Calcium

Principle:

In estimated of calcium the EDTA titrant is used. EDTA solution reacts with both calcium and magnesium. To determine calcium only, the pH is made sufficiently high (by using NaOH buffer) so that the magnesium precipitates as magnesium hydroxide and an indicator (Murexide) specific for calcium is used.

Apparatus:

(*i*) Conical flasks

(*ii*) Burette and Pipette.

Reagents:

(*i*) Sodium hydroxide buffer (1 N):

Dissolve 4 g NaOH in distilled water and when cool, dilute to 100 ml.

(*ii*) Murexide indicator.

(*iii*) Standard EDTA titrant (0.01 M).

Some as in hardness estimation.

(1 ml titrant=l mg $CaCO_3$=0.4008 mg Ca).

Procedure:

(*i*) Take 50 ml of water sample and add 1 to 2 ml NaOH solution to produce a pH 13 to 14.

(*ii*) Add about 0.2 g indicator powder. It gives pink colour to the sample.

(*iii*) Titrate with standard EDTA wish continuously stirring, till the pink colour turns to purple or light blue. This indicates the end point.

(*iv*) Record the vol. of used titrant in ml.

Calculation:

$$\text{mg Ca per litre} = \frac{\text{Used Vol. of titrant} \times A \times B}{\text{Vol. of sample}}$$

where,

A = .4008, (represents that when 1 ml of titrant used .4008 mg of Ca present)

B = 1000 ml.

MAGNESIUM

Magnesium in natural water comes mainly from the leaching of igneous and carbonates rocks. In areas where these sources are common, magnesium concentrations in water often range from 5 to 50 mg per litre. Magnesium is related to water hardness in the same manner as calcium and it is also an essential nutrient for plant growth and development.

Estimation of Magnesium

Procedure :

Magnesium. (Mg^{++})

Chemicals.

EDTA solution, 0.01 N

Dissolve 3.723 gm of EDTA in distilled water to prepare 1000 ml solution.

Buffer solution :

(1) Dissolve 16.9 gm NH_4CI in 143 ml of concentrated NH_4OH.

(2) Dissolve 1.179 gm of EDTA and 0.780 gm $MgSO_4.7H_2O$ in 50 ml distilled water.

Mix both (1) and (2) solutions and dilute to 250 ml with distilled water.

Eriochrome Black T

Mix 0.40 gm Eriochrome Black T, with 100 gm NaCl **and** grind.

Procedure :

(1) Find out the volume of EDTA used in calcium determination.

(2) Also find out the volume of EDTA used in hardness (Ca + Mg).

Calculation :

$$\text{Mg mg/L} = \frac{y - x \times 400.8}{\text{Volume of sample} \times 1.645}$$

where y = EDTA used in hardness determination
x = EDTA used in calcium determination for the same volume of the sample.

SODIUM AND POTASSIUM

Estimation of Sodium and Potassium

(Flame emission spectrophotometry method).

Principle:

Measurement of the light emitted by Na^{++} and K^{+} when these elements are excited in a flame.

(K ; 769 mg ; Na ; 589 mg).

Procedure:

Follow the instructions provided by the manufacture. The samples must be free of solids.

Prepare a calibration curve. This is often appreciably nonlinear. In the case of determination of Na+, dilute the sample until the Na+ content in lower than 5 mg/1.

Interferences:

The nature and extent of interferences depend very much on the characteristics of the flame, and hence on the particular instrument used. Internal standard should therefore be used.

No detailed guidance can therefore be given, but it may be noted that in some cases mutual interference of Na^{+} and K^{+} may be serious. In this case make a K^{+} calibration curve using K^{+} solutions containing the same concentration of Na^{+} is present in the sample and reciprocally for Na^{+}. Anion interferences (particularly PO_4^{3-} and SO_4^{2-}) may also be troublesome.

Note. Collect and store samples in polythylene or polypropylene bottles only.

IRON

Phenanthroline Method

Procedure : *Total Iron.*

— Mix sample thoroughly and measure 50 ml into 9125 ml erlenmeyer flask.

— Add 2 ml conc. HCI and 1 ml $Na_2OH.HCI$ solution. Add few glass beads and heat to boling.

— To insure dissolution of OH the iron continue boiling until volume is reduced to 15 to 20 ml.

— Cool to room temperature and transfer to a 50- or 100 ml volumetric flask or nessler tube.

— Add 10 ml $NH_4C_2H_2O_2$ buffer solution and 4 ml phenanthroline solution, and diluty to mark with water.

— Mix thoroughly and allow at least 10 to 15 min for maxi-' mum colour development.

— Read standards against distilled water set at zero absorbance and plot calibration curve including a blank.

— Selection of light path length for various iron concentration:

Feμg		**Lights path cm**
50 ml Final volume	***100 ml Final volume***	
50-200	100-400	1
25-100	50-200	2
10-40	20-80	5
5-20	10-40	10

APPARATUS

Spectrophotometer ; for use at 510 nm, proving a light , path of 1 cm or longer.

Reagents:

— *HCI.* conc.

— *Hydroxylamine solution.* Dissolve 10 gm $NH_2OH.HCI$ in 100 ml water.

— *Ammonium acetate buffer solution.* Dissolve 250 gm.

NH_4 $C_2H_2O_2$ in 150 ml water. Add 700 ml conc. acetic acid.

— *Phenanthroline solution.* Dissolve 100 mg 1,10 phenanthroline monohydrate $C_{12}H_3N_2.H_2O$ in 100 ml water by stirring and heating to 80°C. Do not boil. Discard the solution if it darkens. Heating is unnecessary if 2 drops conc. HCI are added to the water.

— *Stock iron solution.* Use electrolytic iron wire ; clean wire with fine sand paper to remove any oxide coating and to produce a bright surface.

Weigh 200 mg wire and place in a 1000 ml volumetric flask. Dissolve in 20 ml 6N sulphuric acid and dilute to mark with water,

1.00 ml=200 μg Fe.

— Pipet 50 ml stock solution to 1000 ml Δw.

1 ml= 10 μg Fe.

— Pipet 5 ml stock solution 100 ml

1 ML = 1 μg Fe.

MANGANESE : (PERSULPHATE METHOD)

Chemicals Required

(1) *Special reagent.* Dissolve 75 gm $HgSO_4$ in 400 ml conc. HNO_3 and 200 mL distilled water. Add 2 mL 85% phosphoric acid (H_3PO_4) and 35 mg silver nitrate (Ag NO_3). Dilute the cooled solution in one L.

(2) Ammonium persulphate. $(NH_4)_2S_2O_8$ solid.

(3) Standard manganese solution. Prepare a 0.1 N potassium permagrats ($KmNO_4$) solution by dissolving 3.2 gm in distilled water and making upto 1 L. Age for several weeks in sunlight or heat for several hours near the boiling point, then filter through a fine fritted-glass filter crucible and standardize against sodium oxalate.

$Na_2C_2O_4$. Weigh several 100-to 200 mg samples of $Na_2C_2O_4$

to 0.1 mg and transfer to 400 ml beakers. To each beaker, add 100 ml distilled water and stir to dissolve. Add 10 ml 1 + 1 H_2SO_4 and heat rapidly to 90 to 95°C. Titrate rapidly with the $KmNO_4$ solution to be standardised, *while* stirring, to a slight pink end-point colour that persists for at least 1 min. Do not let temperature fall below 85°C. If necessary, warm beaker contents during titration.

100 mg $Na_2C_2O_4$ will consume about 15 ml permanganate solution. Run a blank on distilled water and H_2SO_4.

$$\text{Normality of } KmNO_4 = \frac{g\ Na_2C_2O_4}{(A - B) \times 0.06701}$$

where A = ml titrant for sample

B = ml titrant for blank.

(4) Standard manganese solution (Alternate). Dissolve 1.000 gm manganese metal (99.8% min) in 10 ml HNO_3. Dilute to 1000 ml with 1 % HCI ; 1 mL = 1.000 mgMn.

Dilute 10 mL to 200 mL with distilled water ; I mL = 0.05 mg Mn. Prepare dilute solution daily.

(5) Hydrogen peroxide. H_2O_3, 30%. (6) Nitric acid, HNO_3 Con.

(7) Sodium nitrate solution. Dissolve 5.0 g $NaNO_2$ in 95 mL distilled water.

(8) Sulphuric Acid. H_2SO_4 conc.

(9) Sodium Oxalate, $Na_2C_2O_4$ primary standard.

(10) Sodium bisulphate. Dissolve 10 gm $NaHSO_3$ in 10 mL distilled water.

Apparatus:

(1) *Spectrophotometer*. For use at 525 nm, providing a light path of 1 cm or longer.

(2) *Filter photometer*. Providing a light path of 1 cm or longer

and equipped with a green fiiter having a maximum transmittance near 525 nm.

(3) *Nessler tubes*. Matched, 100 ml tall form.

Procedure:

To a suitable sample portion add 5 ml special reagent and 1 drop H_2O_2. Concentrated 90 ml by boiling or dilute to 90 ml. Add 1 g $(NH_4)_2S_2O_7$ bring to a boil and boil for 1 min.

Remove from heat source, let stand 1 min, then cool under the tap. Dilute to 100 ml with distilled water and mix.

Prepare standards concentration 0.500...1500 μg Mn by treating various amount of standand Mn solution in the same way.

Use standards prepared, containing 5 to 100 μg Mn/100 ml final volume compare samples and standards visually.

Photometric determination:

Use a series of standards from C0-1500 μg Mn/100 ml final volume. Make photometric measurements against a distilled water blank. The following table shows light path length for various amounts of Magnese in 100 ml volume.

Mn Rang μ_3	*Light path un*
5—200	15
20—400	5
50—1000	2
100—1500	1

Prepare a calibration curve of manganese concentration *Vs* abserbance form the standards and determine Mn in the samples from the curve.

Calculation:

When all the original sample is taken for analysis

$$\text{mg Mn / L} = \frac{\mu\text{g Mn (in 100 ml final values)}}{\text{ml of sample}}$$

HYDROGEN SULPHIDE

Methylene Blue Method

Apparatus:

(a) Matched test tubes approximately 125 mm long and 15 mm OD.

(b) Droppers, delivering 20 drops/ml methylene blue solution. To obtain uniform drops hold dropper in a vertical position and let drops from slowly.

(c) If photometric rather than visual colour determination will be used, either

(i) Spectrophotometer, for use at a wavelength of 664 nm with cells providing light paths of 1 cm and 1 mm, or

(*ii*) Filter photometer, with a filter providing maximum transmittance near 600 nm.

Reagents:

(*i*) Amine-sulphuric acid stock solution. Dissolve 27 g N, N-dimethyl-p- phenylenediamine oxalate in a cold mixture of 50 ml conc. H_2SO_4 and 20 nil distilled water. Cool and dilute to 100 ml with distilled water.

(*ii*) *Amine-sulphuric acid reagent*. Dilute 25 ml amine-sulphuric acid stock solution with 975 ml 1 + 1 H_2SO_4. Store in a dark stars bottle.

(*iii*) *Ferric chloride solution*. Dissolve 100 g $FeC1_3.6H_2O$ i n 40 ml water.

(d) Sulphuric acid solution, H_2SO_4, 1 + 1.

(*e*) *Diammonium hydrogen phosphate solution*. Dissolve 400 g $(NH_4)_2$ HPO_4 in 800 ml distilled water.

(*f*) Methylene blue solution

(*i*) Use USP grade dye or one certified by the Biological stain commission. Dissolve 1.0 g in distilled water and make up to 1L.

0.05 ml (1 drop)= 1.0 mg sulphide/L.

(*ii*) Methylene blue solution II. Dilute 10.00 nil of adjusted methylene blue solution 1 to 100 ml.

Procedure:

(a) Colour Development

1. Transfer 7.5 ml sample to each of two matched test tubes, using a special wide tip pipet or filling to marks on test tubes.
2. Add to tube A 0.5 ml amine-sulphuric acid reagent and 0.15 ml (3 drops) $FeCl_3$ solution.
3. Mix immediately by inverting slowly, only once.
4. To tube B add 0.5 ml 1 + 1 H_2SO_4 and 0.15 ml (3 drops) $FeCl_3$ solution and mix.
5. Colour development-usually is complete in about 1 min, but a longer time often is required for fading out of the initial pink colour.
6. Wait 3 to 5 mix. and add 1.6 ml $(NH_4)_2$ HPO_4 solution to each tube.
7. Wait 3 to 15 min and make colour comparisons.
8. If zinc acetate was used, wait at least 10 min before making a visual colour comparison.

(b) Colour Determination:

(*i*) *Visual colour estimation*. Add methylene blue solution I or II, depending on sulphide concentration and desired accuracy dropwise, to the second tube, until colour matches that developed in first tube. If the concentration exceeds 20 mg/L repeat test with a portion of sample diluted to one fenth.

With methylene blue solution 1. adjusted so that 0.05 ml (I drop) =1.0 mg S^{2-}/L when 7.5 ml of sample are used. mg S^{2-}/L=no. drops solution 1 + 0.1 (no. drops solution II) (*ii*) *Photometric colour measurement*. A cell with a light path of 1 cm is suitable for measuring sulphide concentrations from 0.1 to 2.0 mg/L. Use shorter or longer light paths for higher or lower concentrations. The upper limit of the method is 20 mg/L. Zero instrument with a portion of treated sample from tube B. Prepare calibration curves on basis or colourimetric tests made on Na_2S solutions simultaneously analyzed by the iodometric method, plotting concentration vs. absorbance. A straight-line relationship between concentration and absorbance can be assumed from 0 to 1.0 mgh.

Read sulphide concentration from calibration curve.

FLUORIDE (F^-) SPADNS METHOD

Apparatus:

Spectrophotometer for use at 570 nm providing a light path of at least 1 cm. *Procedure:*

— prepare fluoride standards in the range of 00 to 1.40 mg F/L by diluting appropriate quantities of standard fluoride solution to 50 ml with distilled water.

— Pipet 5.00 ml each of SPADNS solution and Zirconyl-acid reagent, or 10.0 ml mixed acid-Zirconyl SPADNS reagents, to each standard and mix well.

— Set photometer to zero absorbance with the reference solution and obtain absorbance readings of standards. Plot a curve of the milligrams fluoride-absorbance relationship.

1. *Standard Fluoride*

(*a*) Dissolve 221 mg anhydrous NaI in distilled water and dilute to 1000 ml ; 1.00 ml =100 μg LF

(b) Dilute 100 ml stock fluoride solution to 1000 ml with distilled water ;

1.00 ml=10 μgF.

2. *SPADNS Solution:*

Dissolve 958 mg SPADNS in 500 ml distilled water.

3. *Zinconyl-acid reagent:*

Dissolve 133 $ZnOCl_2 . 8H_2O$ in about 25 distilled water. Add 350 ml conc. HCl and dilute to 500 ml with distilled water.

4. *Reference solution:*

Add 10 ml SPADNS solution to 100 ml D.W. Dilute 7 ml conc. HCl to 10 ml and add to the diluted SPADNS solution. The resulting solution, used for setting the instrument reference point (zero).

PHOSPHATE

Phosphorus in waters is present in several soluble and particulate forms, including organically bound phosphorus, inorganic polyphosphates and inorganic orthophosphates. All inorganic phosphate is usually considered as PO_4.

Phosphorus is a biologically active element. Natural source of phosphorus to waters is from leaching of phosphate bearing rocks, and from organic matter decomposition, additional sources are found in man-made fertilizers, domestic sewage and detergents.

Estimation of Phosphate

Apparatus:

(*i*) Spectrophotometer

(*ii*) Acid-washed glassware (flasks & pipettes).

Reagents:

(*i*) Phenolphthale in indicator solution. For preparation see section estimation of alkalinity.

(*ii*) Standard sulphuric acid (.02 N). For preparation see section estimation of alkalinity.

(*iii*) Ammonium molybdate solution (strong acid solution). Dissolve 25.0 g $(NH_4)_6\ MO_7O_{24} . 4H_2O$ in 175 ml distilled water. Continously add 310 ml conc. H_2SO_4 to 400 ml distilled water, cool and add the molybdate solution and dilute to 1 liter.

(*iv*) Stannous chloride solutrion. Dissolve 2.5 g of a fresh supply of $SnCl_2.H_2O$ in 10 ml conc. HCl; then dilute to

100 ml with distilled water. Filter if turbid. (Store in cool).

(*v*) Stock phosphate solution. Dissolve 0.7164 g of potassium dihydrogen phosphate (KH_2PO_4) (which has been dried in an oven at 1059 C for 1 hour) in distilled water. Dilute the soulution to 1 liter 1 ml of stock solution contains 0.5 mg of PO_4.

(*vi*) Standard phosphate solution. Dilute 100.0 ml of stiock solution to 1 liter, with distilled water. 1 ml. of this standard solution contains 0.05 mg of PD_4.

Procedure:

(*i*) Take 50 ml of water sample in a flask and add 2-3 drops of phenolphthale in solution.

(*ii*) If a pink colour appears, add standard sulphuric acid (.02.N) drop by drop until the colour disappears.

(*iii*) Add 2 ml of ammonium molybdate solution with thorough mixing after each addition.

(*iv*) Now .25 ml (5 drops) stannous chloride solution with thorough mixing.

(*v*) After 10 minutes measures absorbance of sample at 690 nm in the spectrophotometer using the reagent blank as gthe reference solution. Determine concentrationof phosphate in teh sample from a standard phosphate calibaration curve.

Calculation :

$$\text{MgPO}_4 \text{ per litre} = \frac{\text{Mg PO}_4 \text{ from curve} \times 1000}{\text{ml of sample}}$$

Standard curve for Phosphate:

(*i*) Take different volumes (these are 1 to 10 ml) of standard phosphate solution in different beakers and dilute (each) to 50 ml with distilled water. Each ml of standard solution contains 0.05 mg of PO_4SO_4 it gives a range of .05 to 0.5 mg of PO_4.

(*ii*) Add 2 ml ammonium molybdate solution and .25 ml stannous chloride solution with thorough mixing, in each sample.

(*iii*) Measure the absorbance of each sample at 90 nm in the spectrophotometer, using the reagents blank as the reference solution.

(*iv*) Draw curve between measured optical densities and known amounts of PO_4 in mg for them.

NITRATE

Nitrate-nitrogen is the most highly oxidized state of the element found in water. In most natural waters it is also the commonest state. It is brought inaquatic body by the bacterial oxidation of atmospheric nitrogen and by decomposition of organic matter in the watershed.

Estimation of Nitrate

(Phenoldisulphonic acid method)

Apparatus:

(*i*) Spectrophotometer

(*ii*) Glassware.

Reagents:

(i) Silver sulphate solution

Dissolve 4.40g $AgSO_4$, free from nitrate in dislilled water and dilute to one litre. (1 ml= 1.0 mg of Cl).

(ii) Phenoldisulphonic acid reagent

Dissolve 25 g pure white phenol in 150 ml conc. H_2SO_4. Add 75 ml fuming H_2SO_4 (15% free SO_3), stir well and heat for two hours on a hot water bath.

(iii) KOH (12N)

Dissolve 73g KOH in distilled water and dilute to 1 litre,

(iv) Ammonium hydroxide concentrated

(v) Stock nitrate solution

Dissolve 0.722 of anhydrous KNO_2 and dilute to 1.0 litre with distilled water. This solution contains 100 mg N per litre.

(vi) Standard nitrate solution

Evaporate 50 ml (which contains 5 mg of N) to dryness on the hot water both dissolve the residue by rubbing with 2 ml phenoldisulphonic acid reagent and dilute to 500 ml with distilled water. 1 mg of this standard solution = 0.01 mg N = 0.0443 mg nitrate ion.

Procedure:

(*i*) Take 50 or 100 ml of water sample and determine the chloride contents of water and treat the sample with an equivalent amount of standard silver sulphate solution. Remove the precipited chloride either by centrifugation or by filtration, coagulating the silver chloride by heat if necessary.

(*ii*) Evaporate the solution to dryness over a hot water both-rub the recidue thoroughly with 2 ml phenoldisulphonic acid reagent to insure solution of all solids. If need be heat on hot water both for a short time to dissolve entire residue. Dilute with 20 ml distilled water.

(*iii*) Add with stirring about 6 to 7 ml conc. NH_4OH or about 5 to 6 ml 12 N KOH until maximum colour developed. Filter any resulting flocculent hydroxide from the coloured solution. Remove the prescipited chloride either by centrifugation or by filtration, coagulating the silver chloride by heat if necessary.

(*iv*) Evaporate the solution to dryness over a hot water both-rub the recidue thoroughly with 2 ml phenoldisulphonic acid reagent to insure solution of all solids. if need be heat on hot water both for a short time to dissolve entire residue. Dilute with 20 ml distilled water,

(*v*) Add with stirring about 6 to 7 ml conc. NH_4OH or about 5 to 6 ml 12 N KOH until maximum colour developed. Filter any resulting flocculent hydroxide from the coloured solution. Transfer the filtrate to a 50 or 100 ml volumetric flask, dilute to the mark and mix.

NITROGEN (NITRATE) (COLORIMETRIC METHOD)

Apparatus:

1. *Spectrophotometer*, for use at 543 nm, providing a light path of 1 cm or longer.
2. Filter photometer, providing a light-path of 1 cm or longer and equipped with a green filter having maximum transmittance near 540 nm.

Reagents:

(*a*) Nitrite Free Water.

(*b*) Colour reagent. To 8.0 ml water add 100 ml 85% phosphoric acid and 10g sulphanilamide. After dissolving sulphanilamide completely, add 1g. N-(1-naphthyl)-ethylenediamine dihydro-chloride. Mix to dissolve, then dilute to I L with water. Solution is stable for about a month when stored in a dark bottle in refrigerator.

(*c*) Sodium oxalate 0.025 M (0.05 N). Dissolve 3.350 g $Na_2C_2O_4$, primary standard grade, in water and dilute to 1000 ml.

(*d*) Ferrous ammonium sulphate, 0.05 M (0.05 N). Dissolve 19.607 g $Fe(NH_4)_2\ (SO_4)_2 .\ 6H_2O$ plus 20m.1 conc. H_2SO_4 in water and dilute to 1000 ml.

(*e*) Stock nitrite solution. Dissolve 1.232 g $NaNO_2$ in water and dilute to 1000 ml 1.00 ml = 250 μg N. Preserve with 1 ml $CHCI_3$.

(*f*) Intermediate nitrite solution. Calculate the volume, G, of stock NO_2^- solution required for the intermediate NO_2^- solution from G=12.5/A. Dilute the volume G (approximately 50 ml) to 250 ml with water, 1.00 ml = 50.0 μg N. Prepare daily.

(g) Standard nitrite solution. Dilute I0.C0 ml intermediate NO_2^- solution to 1000 ml with water ; 1.00 ml = 0.500

μg N. prepare daily.

Procedure:

(*a*) Removal of suspended solids. If sample contains suspended solids, filter through a 0.45 μm pore diam membrane filter.

(b) Colour development. It sample pH is not between 5 and 9 ; adjust to that range with 1 N HCl or NH_4OH as required. To 50.0 ml sample, or to a portion diluted to 50.0 ml, add 2 ml colour reagent and mix.

(c) Photometric measurement. Between 10 min and 24 after adding colour reagent to samples and standards, measure absorbance at 543 nm.

Calculation:

Prepare a standard curve by plotting absorbance of standard against NO_3^-N concentration. Compute sample concentration directly from curve.

NITROGEN-AMMONIA

Phenate Method

Apparatus:

(i) Colorimetric equipment. One of the following is required:

(*a*) Spectrophotometer, for use at 600 nm with a light path of approximately 1 cm.

(b) Filter photometer. Equipped with a red orange filter having a maximum transmittance near 630 nm and providing a light path of approximately 1 cm.

(*ii*) Magnetic stirrer.

Reagent:

(*a*) Ammonia-Free Water. Prepare by ion exchange or distillation methods.

(*b*) Hypochlorous acid reagent. To 40 ml water add 10 ml 5% NaOCI solution prepared from commercial bleach. Adjust pH to 6.5 to 7.0 with HCI. Prepare this unstable reagent weekly.

(*c*) Manganous sulphate solution 0.003 M. Dissolve 50 mg $MnSO_4.H_2O$ in 100 ml water.

(*d*) Phenate reagent. Dissolve 2.5g NaOH and 10 g phenol. $C_6H_2OH_5$ in 100 ml water. Because this reagent darkens on standing, prepare weekly.

(*e*) Stock ammonium solution. Dissolve 381.9 mg anhydrous NHCI dried at 100°C, in water, and dilute to 1000 ml.

$$1.00 \text{ ml} = 100\ \mu g \text{ N} = 122\ \mu g \text{ NH}_3$$

(*f*) Standard ammonium solution. Dilute 5.00 ml stock ammonium solution to 1000 ml with water. 1.00 mL = 0.500 µg N = 0.607 µg NH_3

Procedure:

(a) Treatment of sample:

1. To a 10.0 ml sample in a 50 ml beaker, add I drop (0.05 ml) $MnSO_4$ solution.
2. Place on a magnetic stirrer and add 0.5 ml hypochlorous acid reagent.
3. Immediately add a drop at a time, 0.6 mL phenate reagent.
4. Add reagent without delay using a bulb pipet or a buret for convenient delivery.
5. Stir vigorously during addition of reagents. Because colour intensity is affected by age of reagents ; carry a blank and a standard through the procedure with each batch of samples.
6. Measure absorbance using reagent blank to zero the spectrophotometer.
7. Colour formation is complete in 10 min. and is stable forat ieast 24 h.
8. Although the blue colour has a maximum absorbance at 630 nm satisfactory measurements can be made in the 600 to 660 nm region.

(b) Preparation of Standards. Prepare a calibration curve in the NH_3 -N range of 0.1 to 5 µg, treating standard exactly as the sample. Beer's law governs.

Calculation:

Calculate ammonia concentration as follows mg NH_3–N/L (11.1 ml final volume)

$$= \frac{A \times B}{C \times S} \times \frac{D}{E}$$

where

A = absorbance of sample

B = NH_3–N is standard, µg

C = absorbance of standard

S = volume of sample used, ml

D = volume of total distillate collected, ml, including acid aborbant, neutralizing agent, and ammonia free water added

E = volume of distillate used for colour development, ml.

The ratio D/E applies only to distilled samples.

Calculation:

$$\text{Mg of N per litre} = \frac{\text{mg N from standard curve} \times 1000}{\text{ml of sample}}$$

$$\text{mg of } NO_3 \text{ per litre} = \text{mg of N per litre} \times 4.43$$

Standard curve for Nitrate:

(*i*) Take different volumes of standard nitrate solution (these are 1 to 10 ml) in different beakers and dilute to 50 ml (each sample). The standard nitrate solution contains .01 mg N per ml or 0.0443 mg nitrate ion, so it gives a range of .01 to 0.1 mg N.

(*ii*) Estimate nitrogen nitrate by above described procedure in each sample.

(*iii*) Measure the absorbance of each sample at 400-425 nm in the spectrophotometer, using the reagent blank as the reference solution.

(*iv*) Draw curve between measured optical densities and known amount of N or nitrate for them.

SILICATE

Silicate is extremely common in nature as a constituent of ignous rocks, quartz and sand. Natural waters often contain 1 to 10 mg silica per litre and very rarely exceed 60 mg. silica per litre.

Estimation of Silica

Apparatus:

(*i*) Spectrophotometer

(*ii*) Flasks and pipettes etc.

Reagents:

(*i*) Hydrochloric acid 0.25 N.

Dilute 22 ml conc. HCl (sp. gr. 1.19) to 1 litre with distilled water.

(*ii*) Ammonium molybdate solution (5%).

Dissolve 52g of ammonium molybdate in 1 litre of distilled water.

(*iii*) Na_2 DTA (10%)

Dissolve 10g Na_2 EDTA in 1 litre distilled water.

(*iv*) Sodium sulphite (17%)

Dissolve 170g Na_2SO_3 in 1 litre distilled water.

(*v*) Stock sodium silicate solution

Dissolve 1.765g of $Na_2SiO_3 . 5H_2O$ in water and dilute to 1 litre. (1 ml=0.5 mg SiO_3).

(*vi*) Standard sodium silicate solution

Dilute 100 ml of stock sodium silicate solution to 1 litre. 1 ml of standard solution contain 0.05 No.SiO_3.

Procedure:

(*i*) Take 10 ml of sample and add 5 ml 0.25 NHCI.

(*ii*) Add 5 ml of 5% Ammonium molybdate.

(*iii*) Add 5 ml 10% Na_2 EDTA.

(*iv*) After five minutes, add 10 ml Na_2SO_3.

(*v*) Now allow to stand for 30 minutes to develop colour.

(*vi*) Determine absorbance at 7O0 nm in spectrophotometer, using reagent blank as the reference solution.

(*vii*) Determine concentration of silica from the standard calibration curve.

Calculation:

$$\text{mg } SiO_2 \text{ per litere} = \frac{\text{mg of } SiO_2 \text{ from curve} \times 1000}{\text{ml of sample}}$$

Standard curve for Silica

(*i*) Take different volumes (these are 1 to 10 ml) of standard sodium silicate solution in different beakers and dilute (each sample) to 50 ml with distilled water. This gives a range of 0.05 to 0.5 mg SiO_3, because 1 ml of standard sodium silicate solution contains .05 of $Si0_3$.

(ii) Now estimates silica in all samples by above given procedure.

(*iii*) Measure the absorbance of each sample at 700 nm in the spectrophotometer, using reagents blank as the reference solution.

(*iv*) Draw curve between measured optical densities and known amounts of SiO_2 in mg for them.

SULPHATE

In many natural water, sulphate is the second most common anion, being derived from most sedimentary rocks. In lakes, sulphur is cyclic and involves organically reduced forms as well as the common free SO_4^- ion.

Estimation of Sulphate (By Turbidometric Method)

Apparatus:

(*i*) Spectrophotometer

(*ii*) Glassware.

Reagents:

(*i*) Acid salt solution

Dissolve 240g NaCl in approximately 900 ml distilled water. Add 20 ml conc. HCl and dilute to 1 litre.

(*ii*) Barium Chloride crystals

20 to 30 mesh.

(iii) Standard sulphuric acid solution : (0.02 N)

1 ml diluted to 50 ml will give a solution of 19.2 mg/l.

Principle:

Sulphate froath is prepared with barium chloride in acid sodium in such manner as to form barium sulphate crystals of uniform size. Light transmitted by the turbid solution is measured with a photo-electric colourimeter and sulphate ion concentration is read from a standard curve. (Colorimetric equipment 420 nm).

Procedure:

(*i*) Measure 50 ml sample into a 250 ml flask. If the sample contains more than 60 mg/I sulphate, an oliqu of portion should be taken and distilled water added to make 50 ml. Add exactly 10 ml acid salt solution and mix. If the sample is not perfectly clear and colourless at this point, the apparent sulphate content should be measured photometrically and this value substracted from the determined sulphate concentration.

(*ii*) Add one spoon full $BaCl_22H_2O$ crystals and start magnetic stirrer with delay the time of stirring should be constant within ±2 Sec. although the total time interval may be any convenient time from 45 Sec. to 3 min. One minute is very satisfactory.

(*iii*) Allow to stand for an additional 4 min. exactly and transfer the resultant suspension to the colourimeter and read immediately.

Then read sulphates concentration on a curve prepared from solutions of known sulphate concentrations.

(*iv*) Standard 0.02 N sulphuric acid is a convenient solution to use in varying dilutions for preparing the standard curve, 1 ml diluted to 50 ml will give a solution of 19.2 mg/l. The standard curve is not a straight line on semilogarithmic paper, but generally will have a point of inflection so that it should be prepared from at least 10 dilutions ranging from 5 to 0 mg/1 sulphate.

Calculation:

The calibration curve may be plotted in term of SO_4 concentration based on a 50 ml sample, versus per cent absorbance. The concentration is then read from the curve. If a different amount of sample was taken, multiply by the appropriate factor.

$$\text{mg } SO_4 \text{ per litre} = \frac{\text{mg } SO_4 \text{ in sample} \times 1{,}000}{\text{ml of sample}}$$

$$= \text{mg } SO_4 \text{ in sample} \times 20$$

TOTAL SOLIDS OR RESIDUES

Residue refers to material left in a versel after evaporation of a water sample. It may be sub-divided into organic matter and inorganic material. These may both be further sub-divided into particulate and dissolved matter. Total residue can be determined by simply the weight of material remaining after evaporation of raw water.

Determination of Total Residue

Equipment:

(*i*) Evaporating dishes.

(*ii*) Drying even at 103°C.

(*iii*) Analytical balance.

(*iv*) Desiccator etc.

Procedure:

(*i*) Collect sample in Pyrex or polyethylene bottle.

(*ii*) Determine the weight of empty dish. (The empty dish must be taken through the same series of steps to determine the weight, as for determination of weight of dish with residual).

(*iii*) To the prepare and prcweighed dish add a well mixed' aliquot of water sample to be evaporated. Usually 100 ml will suffice, but exact volume will be determined by the sensitivity of the balance. It must not contain more than 200 mg of residue.

(*iv*) Evaporate to dryness and then dry overnight in an over at 103°C.

(*v*) Remove dish from oven, cool briefly in the air, and place in desiccator; dishes must not touch to the sides of the dessiccator.

(*vi*) Take desicator to balance, remove cooled dishes and weigh rapidly.

Calculation:

$$\text{mg total residue on drying at } 103°\text{C} = \frac{\text{A} \times \text{B}}{\text{ml of sample}}$$

where,

A = mg total residue

B = 1000 ml.

DISSOLVED SOLIDS OR FILTERABLE RESIDUE

Dissolved solids or filterable residue is the weight or material remaining after evaporation of the filtrate of the raw water sample.

Determination of Dissolved Solids

Equipment:

As for total residue, plus a filter apparatus using, acidwashed ashless, hard finish filter paper for fine precipitates or for most precise work a filter apparatus using 0.45 um membrane filters.

Procedure:

The procedure is the same as for total residue except that the sample is filtered and the filtrate evaporated. Results are reported as filterable residue or total dissolved solids on drying at 103°C.

VOLATILE MATTER

Volatile matter provides only a crude estimate of organic matter present in the water sample mainly because ignition also produce a decomposition or volatilization of inorganic salts. Better estimates of organic matter, especially in low concentrations, may be obtained by quantitative dichromate oxidation or infrared carbon analysis. However, the volatile matter procedure is more simple because the sample used in determination of total or filterable residue is carried on through only one more step.

Equipment

As for residue, but vessel must be able to withstand higher temperatures and must be preignited before initial weight is determined, also, an electric muffle furnace is required.

Procedure:

(*i*) Preheat muffle furnace to 600°C.

(*ii*) Ignite sample used in residue determination for 15 .minutes, air cool briefly, complete cooling in desiccator, and weigh. The lose on ignition is reported as milligrams of volatile solids per litre.

Calculation:

$$\text{mg volatile solids per litre} = \frac{\text{mg volatile solids} \times A}{\text{ml. of sample}}$$

where,

A= 1000 ml.

4

Biological Factors

COLLECTION AND EXAMINATION OF WATER SAMPLES

Biological methods used for assessing water quality include the collection, counting and identification of aquatic organisms; biomass measurements, measurements of metabolic activity rates, measurements of the toxi city, bio-concentration and bioaccumulation of pollutants, and processing and interpretation of biological data. The following communities of aquatic organisms are considered in specific sections that follows: Plankton, a community of plants (*Phytoplankton*) and animals (*Zooplankton*), usually swimming or suspended in water, nonmotile or insufficiently motile to overcome transport by currents. *Periphyton* (Aufwuchs) a community of microscopic plants and animals associated with the surfaces of submersed objects. *Macrophyton*, the larger plants of all types. *Macroinvertebrates*, they are generally bottom-dwelling organisms (*benthos*). The methods selected are necessary for the appraisal of water quality. Principal emphasis is on methods and equipment rather than on interpretation or application of results.

Plankton

Plankton, particularly phytoplankton, long have been used as indicators of water quality. Some species flourish in highly eutropic waters while others are very sensitive to organic and/ or chemical wastes. The species assemblage of phytoplankton and zooplankton also may be useful in assessing water quality.

Sample collection

The frequency and location of sampling is dictated by the purpose of the study. Locate sampling stations as near as possible to those selected for chemical and bacteriological sampling to insure maximum correlation of findings.

Establish a sufficient number of stations in as many locations as necessary to define adequately the kinds and quantities of plankton in the water studied. The physical nature of the water (standing, flowing or tidal) will influence greatly the selection of sampling stations. Samples usually are referred to as 'surface' or 'depth' (subsurface) samples. Sampling frequency depends on the intent of the study as well as the range of seasonal fluctuations, the immediate meteorological conditions, adequacy of equipment, and availability of personnel.

Sampling procedure

Once sampling locations, depths, and frequency have been determined, prepare for field sampling. Label sample containers with sufficient information to avoid confusion or error. If samples are to be preserved immediately after collection, add preserved immediately after collection, and preservative to container before sampling. In a field record book note sample location, depth, type, time, meteorological conditions, turbidity, water temperature, salinity and other significance observations. Collect coin adent samples for chemical analysis to help define environmently variations having a potential effect on plankton.

(*a*) *Phytoplankton.* In oligotrophic waters or where phytoplankton sensities are expected to be low collect a sample of upto 6 L. For richer entrophic waters collect a sample of 0.5 to 1 L. For qualitative and quantitative evaluations collect whole (unfiltered and unsustained) water samples with a water collection bottle consisting of a cylindrical tube with stoppers at each end and a closing device. Lower the open sampler to the desired depth and close by dropping a weight, called a messenger, which slides down the supporting wire or cord and trips the closing mechanism. The most commonly used samples that operate on his principle are the Kemmeres, Van Dorn, Niskin and Nansen samplers. Different size categories of phytoplankton can be separated in the field of laboratory by filtering through neting of the appropriate mesh size. Typically, for net plankton use a No. 20 net with mesh openings of 76 to 80 long (silk bolting cloth or nylon monofilament screen cloth). Select appropriate mesh size for concentrating the various size categories of phytoplankton typical of the aquatic system under study.The most suitable phytoplankton preservative is ingol's solution, other commonly used preservative include Formalin, Merthiolate, glutaraldehyde, 95% alcohol etc.

(*b*) *Zooplankton.* The choice of sampler depends on the type of Zooplankton, the kind of study (distribution, productivity etc.) and the body of water being investigated. For collecting microzooplankton (20 to 200 long) such as protozoa, rotifera and immature microinestacea, use the bottle samplers described for photoplankton. The small zooplankters usually are sufficiently abundant to yield adequate samples in 5 to 10 L bottles, however, composite sampler over depth and time are recommended. Water bottle samplers are suitable especially for discrete depth samples. The larger and more robust microzooplankters (eng. loricate forms and crustacea) may be concentrated by passing the whole water through a 20 long mesh net. If quantitative estimates of the

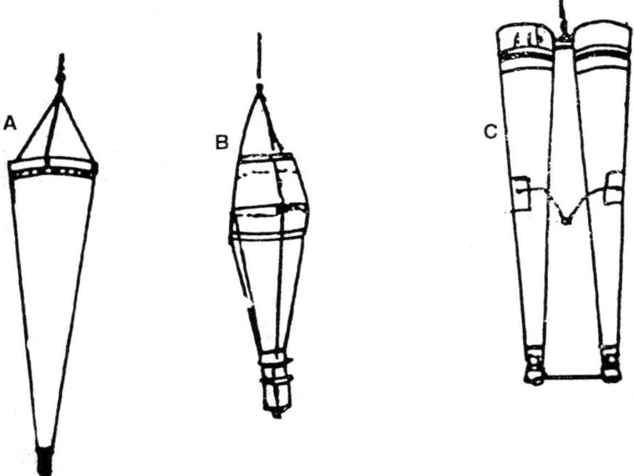

Figure 4.1 : ***Commonly used plankton nets.***

nonloricate, delicate forms are required, do not screen. Fix 0.5 to 5 L of whole water for enumeration of these forms. Bottle samplers usually are unsuitable for collecting large zooplankton, such as nature microinstacea, that unlike the smaller forms, are much less numerous and are sufficiently agile to avoid capture. Although comparatively larger water volumes, and consequently adequate numbers of microinstacea, can be sampled with a pump, avoidance by larger more agile zooplankters at the pump head can cause sampling error. Consequently larger trap samplers (In day trap, Sohindier Patalas trap) or nets like Plankton Tow Nets, Wisconsin Nets, Bongo nets are the preferred collection methods. Table 4.1, denotes the various mesh sizes of plankton Nets.

TABLE 4.1 : CHARACTERISTICS OF COMMONLY USED PLANKTON NETS

Silk No.	*Size of Aperture µm*	*Approx. Open. Area %*	*Classification*
000	1024	58	Largest zooplankton and ichthyoplankton
00	752	54	Larger plankton and ichthyoplankton
0	569	50	Large zooplankton and ichthyoplankton
2	366	46	Large microcrustacea
6	239	44	Microcrustacea
10	158	45	Microcrustacea and most rotifers
20	76	45	Net phyto-and zooplankton
25	64	33	Nannoplankton

Preserve Zooplankton samples with 70% ethanol or 50% buffered formalin. Ethanol preservative is preferred for materials to be stained in permanent mounts or stored. Formalin may be used for the first 484 of preservation with subsequent transfer to 70% ethanol.

Concentration techniques

The organisms contained in water samples sometimes must concentrated in the laboratory before analysis. Three techniques for concentrating phytoplankton, namely, sedimentation, membrane filtration, certifugation and a special technique for Zooplankton are described below.

(*i*) *Sedimentation. Sedimentation* is the preferred method of concentration because it is nonselective (unlike filtration) and nondestructive (unlike filtration or centrifugation), although many of the picoplankton, the smaller nanoplankton, and actively swimming flagellates (in preserved samples) may not settle completely. Allow 14 settling/mm of column depth. For a treated sample (10 ml detergent/L) allow about 0.54 settling/mm depth. The sample may be concentrated in a series of steps by quantitatively transferring the sediment from the initial container to sequentially smaller ones.

(*ii*) *Membrane filtration.* The filtration method permits use of high magnification for enumerating small plankters including flagellates and cyanobacteria. Pour a measured volume of well mixed sample into a funnel Equipped with a membrane filter having a pore diameter of 0.45 cm. Apply a vaccum of less than 50 KPa to the tilter until about 0.5 cm. of sample remains of on filter. Break vaccum, they apply low vaccum (about 12 K Pa) to remove remaining water but not to dry the filter.

(*iii*) *Centrifugation.* Plankton can be concentrated by batch or continuous centrifugation. Centrifuge batch samples at 1000 g for 20 min.

(*iv*) *Zooplankton concentration.* Zooplankton samples often need to be concentrated in the filed especially when a larger water bottle or pump methods of sampling are used. When only small volume reduction are needed, pour sample back into the bucket of traps or nets. In processing large volumes of water as with pump sampling, use larger plankton buckets or funnels with greater water volume retention and filteration surface area.

Preparing slide mounts

(*i*) *Phytoplankton Semi-Permanent-Wet-Mounts-Agitate* the settled sample concentrate and withdraw a subsample with an accurately calibrated automatic pipet. Clean and calibrate the pipet regularly. To prepare wet mounts transfer 0.1 ml to a glass slide, place a cover slip over the sample, and ring the cover slip with an adhesive such as clear nail polish to prevent evaporation. For semipermanent slides, mix glycerin with the sample, as the sample ages the water evaporates, leaving the organisms imbedded in the glycerin.

(*ii*) *Phytoplankton permanent mounts.* (*a*) *Membrane filter mounts* : Place two drops of immersion oil on a labelled slide. Immediately after filtering place the filter on top of the oil with a pair of forceps and add two drops of oil on top of the filter. The oil impregnates the filter and makes it transparent. Impregnation time is 244 to 484. This procedure can be completed in 1 to 24 by applying beat (70°C). Once the filter has cleared, place a fene additional drops of oil on it and cover with a cover slips. The mounted filters is now ready for microscopic examine. (*b*) *Sedimented slide mounts* : *Two* techniques are available for making permanent, resin mounts of natural phytoplankton that has been deposited by sedimentation on a microscope slide or cover glass and dehydrated by ethanol vapour substitution.

(*iii*) *Diatom mounts.* Samples concentrated for diatom analysis by settling or centrifugation may contain dissolved materials, such as marine salts, formalin and detergents, that will leave interfering residues. Wash well with distilled water before slide preparation. Treat samples concentrated for diatom analysis by membranes filteration as described by Patrick and Reimer. Mix equal volumes of conc. nitric acid (HNO_3) and sample Add a few grains of Potassium dicromate ($K_2Cr_2O_7$) to facilitate digestion of the filter and cellular organic matter. Add more dichromate if solution colour changes from yellow to green. Place sample on a hot plate and boil down to approximately one third the original volume. Alternatively, let treated sample stand overnight. This cleaning process destroys organic matter and leaves only diatom shell (frustules) cool, wash with distilled water, and mount as described above. Place a drop of mounting medium in the centre of a labelled slide. Using a suitable microscopic mounting medium assured permanent easily handled mounts for examination under oil immersion. Heat the slide to near 90°C for 1 to 2 min before applying the heated cover slip with its sample residue to hasten evaporation of solvent in the mounting medium.

(*iv*) *Zooplankton mounts*. For Zooplankton analysis withdraw a 5 ml subsample from the concentrate and dilute or concentrate further as necessary. Transfer sample to a counting cell or chamber for analysis as a wet mount. Use polyvinyl lactylphenol for preparing semi-permanent Zooplankton mounts.

Periphyton

Microorganisms growing on stones, sticks, aquatic macrophytes, and other submerged surfaces are useful in assessing the effects of pollutants on lakes streams and estuaries. Included in this group of organisms were designated periphyton, are the zoogleas and filaments bacteria, attached protozoa, rotifers and algue and the freelining microorganisms that swim, creep or fodge among the attached forms.

Sample collection:

(*a*) *Natural substrates:* Collect qualitative samples by scraping submerged stones, sticks, pilings, and other available substrates. Many devices have been developed to collect quantitative samples from irregular surfaces but success rarely is achieved.

(*b*) *Artificial substrates:* The most widely used artificial substrate is the standard plain, 25 by 75 mm glass microscope slide but other materials such as clear vinyl plastic also are suitable. Do not change substrate type during a study because colonization varies with substrate.

(*c*) *Exposure period :* Colonization on clean slides proceeds at an exponential rate for the first 1 or 2 week. Because exposures of less than 2 weeks may result in very spare collections, and exposures of more than 2 weeks may result loss of material due to slougjing, 2 weeks usually constitutes the optimum sampling interval during the summer.

Sample preservation

Preserve samples that are taken for counting and identification in 5% neutralized formalin or merthionate.

Sample analysis:

1. *Sedgwick Rafter Counts*: Remove periphyton from slides with a razor blade and rubber policeman. Disperse scrapings in 100 ml or other suitable volume of preservative with vigorous shaking or use a blender. Transfer a 1 ml portion to a sedgwick Rafter cell and make a strip count as described earlier.

Express the counts as cells or filaments persquare millimeter of substrate area, calculated as follows

$$(1)\quad \text{cells / ml suspended scrapings} = \frac{\text{Actual Count / strip}}{\text{Volume of 1 strip, ml.}}$$

$$(2)\quad \text{cells / mm}^2 \text{ slide surface} = \text{cells / ml suspended scrapings} \times \frac{\text{Total volume of scrapings}}{\text{Area of slide or slides mm}^2}$$

2. *Inverted microscope method counts*. Using an inverted microscope for periphyton counts permits magnifications higher than those possible with the *Sedwick-Rafter Cell*. Prepare scrapings as in Sedwick-Rafter Cell, and transfer a measured portion, after serial dilution if necessary into a standardized plankton sedimentation chamber. After a suitable period of setting count organisms as the setting chamber by counting all organisms within a known number of strips or random fields. Calculate algal density per unit area of substrate as follows

$$\text{Organisms / mm}^2 = \frac{N \times A_i \times V_i}{A_c \times V_s \times A_s}$$

where

N = Number of organisms counted.

A_i =Total area of chamber bottom, mm^2.

V_i = Total volume of original sample suspension, ml.

A_c = Area counted (strips or fields), mm^2.

A_s = Sample Volume used in chamber, ml, and

A_s = Surface area of slide or substrate, mm^2.

Benthic Macroinvertebrates

Benthic macroinvertebrates are animals inhabiting the substratum of lakes, streams, estuaries, and marine waters. They may construct attached cases, tubes or nets that they live on or in, or freely over rocks, organic debris, and other substrates during all or past of their life cycle. The composition and density chamber of individuals per unit area) of rnacroinvertebrate communities in streams, lakes, estuaries, and marine waters are reasonably stable from year to year in unperturbed environment. However, seasonal fluctuations associated with life-cycle dynamics of individual species many result in extreme variation at specific sites within an calender year. Macroinvertebrate community responses to environmental perturbations are useful in assessing the impact of municipal, industrial oil, and agricultural wastes, and impacts from other land used on surface water bodies. Assessing the impact of a pollution source generally involves comparison of macro invertebrates communities and their habitates at sites influenced by pollution with those collected from adjacent unaffected sites.

Sample collection

Quantitative and qualitative samplers have been designed to collect organisms from stream and lake bottoms. The mot common quantitative sampling devices are the ponas, pekrson and Ekman grabs and the surfer or square foot stream bottem sampler.

(a) Grab sampler:

(*i*) *The Ponar grab*. The ponar grab is used increasingly in medium to deep rivers, lakes and reservoirs. It is similar to the Petersen grab in size, weight, lever system and sample compartment, but has side plates and a screen on top of the sample compartment to prevent sample loss during closure. This sampler is best used for sand, gravel, or small rocks with mud but it can be used in all substrates except bed rock.

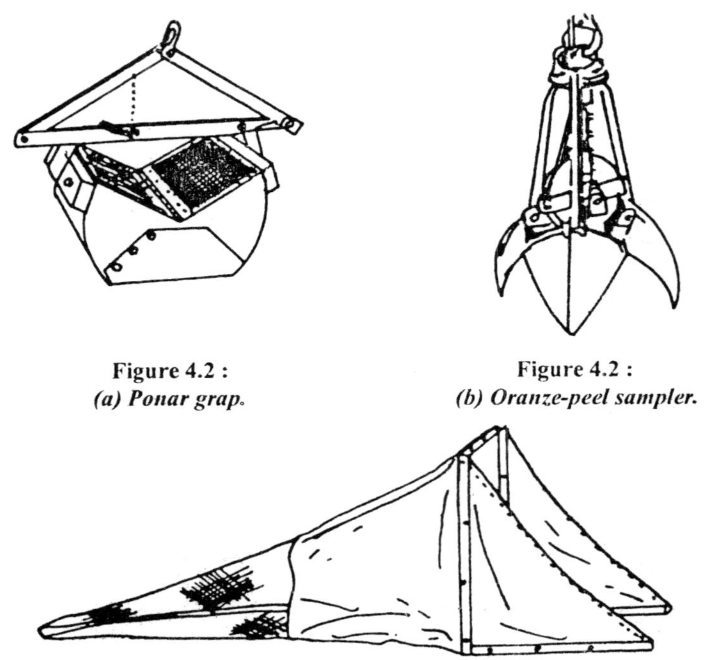

Figure 4.2 : ***(a) Ponar grap.***

Figure 4.2 : ***(b) Oranze-peel sampler.***

Figure 4.2 : ***(c) Surber or square-foot sampler.***

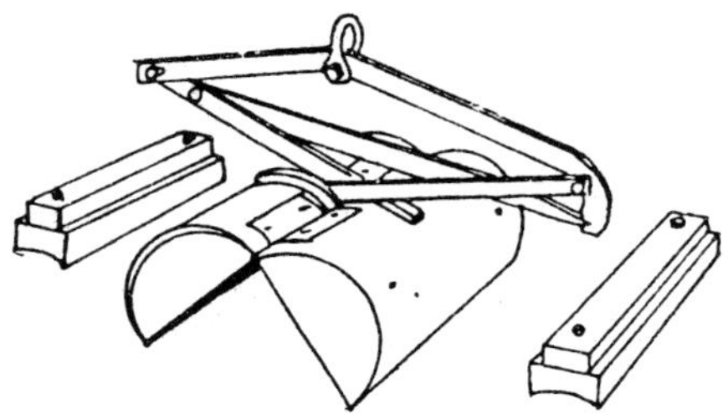

Figure 4.2 : ***(d) Petersen grab.***

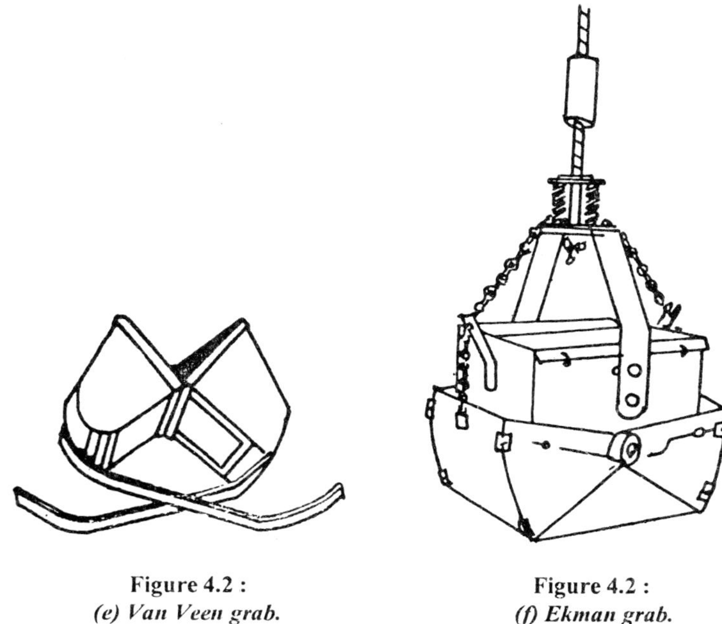

Figure 4.2 :
(e) Van Veen grab.

Figure 4.2 :
(f) Ekman grab.

Figure 4.2 : ***Common sampling devices.***

(*ii*) *The Orange Peel grab.* The orange peel grab is suited for use in marine environments and deep lakes with sandy substrates.

(*iii*) *The Peterson grab.* The peterson grab is used widely for sampling hard bottoms such as sand, gravel, men 1, and cley in scrift-currents and deep water. It is an iron clamtype grab manufactured in various size that will sample an area of from 0.06 to 0.09 m^2. It is useful approximately 13.7 kg. but may weight as much as 31.8 kg. when auxiliary weights are bolted to its sides. The primary advantage of the extra weights is to make the grab stable in swift current and to give additional cutting force in fibrous or firm bottom materials. Modify the sampler by adding end plates, by cutting large strips out at the top of each side, and by adding a hinged 30 mesh screen. To use the Petersen grab, set the hinged jaws and lower to the bottom slowly to avoid disturbing lighter bottom materials. Ease rope tension to release the catch. As the grab is raised the level system closes the jaws.

(*b*) *The Van Veen grab.* The Van Veen grab is useful in sampling in the open sea and in large lakes. The long arm tend to stabilize the sampler without disturbing the water at the water substrate interface. It is basically on improved various of the Petersen grab and is useful on substrates of mud, gravel, pebbles and sand. The sampler is heavy, lower it from a boat or strip platform with mechanical or hydraulic lifts.

The Smith McIntyre grab. The Smith McIntyre grab has the heavy steel construction of the Petersen, but its jaws are closed by strong coil springs.

(*c*) *The Ekman grab.* The Ekman grab is useful only for sampling silt, much, and sludge in water with little current.

It is difficult to use when rocky or sandy bottoms are present because small pebbles or grit prevent proper jaws closure. The grab is made of 1 2 to 20 gauge brass or stain less steel and weighs approximately 3.2 kg. The box like part holding the sample has spring operated jaws on the bottom that must be cocked manually. At the top of the grab are two hinged overlapping lid, that are held open partially during descent by water passing through the sample compartment. These lids are held shut by water pressure when the sampler is being retrieved. The grab is made in three size : 15 × 15 cm ; 23 × 23 cm and 30 × 30 cm, but the smallest size usually is adequate and is most desirable for replicate sampling. To prevent sample overflow and loss, place a standard U.S. No. 30 sieve in sort in the top for deep sediments.

Sample Processing and Analysis

After collecting a bottom grab sample, transfer it to either specially designed sieve tables (or hoppers) or a container. If container is used, dilute with ambient water, and swirl. Pour slurry gradually into a sieve bucket. Gently wash slurry over screen to prevent damaging or losing specimens slurries that alog the screen require removed of screened material. Wash residual on the screen into a container and fix the contents in a solution of 10% buffered formalin or 70% ethanol.

Whether organisms are sorted in the field or the laboratory, follow consistent procedure. Before processing a sample, transfer information from the label to a data sheet that provides space for scientific names and number of individuals. Identify animals in each vial using a stereoscopic and compound microscope, according to need, and available experience and resources. Identify organisms to species level if possible.

Data Evaluation and Presentation

There are two basic approaches used in evaluating effects of pollutants on aquatic life. The first is to make a qualitative analysis of fanra and flora 'above and below' or 'before and after' thereby determining species present or absent. Then through an understanding of the responses of various species to specific pollutants, determine the significance of damage or change.

The second approach is to make a quantitative inventory of the number *of* specimens, species, and structure of the aquatic community affected by the pollutant and to compare with reference information. In most pollution surveys these approaches, are integrated because each provides valuable interpretative information.

Data Presentation

Data presentation may take many forms. The basic techniques including tables, bar graphs (horizontal and vertical), pie diagrams, pictorial charts (ideographs) line graphs, frequency distribution tables and graphs, histograms, frequency polygons, and cummulative frequency polygons. These may be superimposed on maps.

MICROMETRIC METHODS

Compound Microscope

Use either a standard or an inverted *compound microscope* for algal identification and enumeration. Equip either type with a mechanical stage capable *of* moving all parts of a counting cell past the objective lens. Standard equipment is a set of 10 × or 12.5 × oculars and 1 0 ×, 20 ×, 40 ×, and 100 × objectives. Use objectives to provide adequate working distance for the counting chamber. Magnification requirements vary with the plankton fraction being investigated, the type of microscope, counting chamber used, and optics. The SedgwickRafter chamber limits mangification to approximately 200 ×. The Palmer-Maloney cell permits magnification up to 500 ×. Inverted microscopes are limited in resolution by their optics. The useful upper limit of magnification for any objective is 1000 times the numerical aperture (NA). Above this magnification, no greater detail can be resolved. Use combinations of oculars, intermediate magnifiers, and objectives to obtain the greatest magnification without exceeding the useful limit of magnification. When the limit is exceeded, empty magnification results. Empty magnification occurs where the image is larger but no greater resolution is achieved. Always maximize resolution. Optics providing contrast enhancement such as phase contrast or differential interference contrast are useful.

Stereoscopic Microscope

The *stereoscopic microscope* is essentially two complete microscopes assembled into a binocular instrument to give a stereoscopic view and an erect rather than an inverted image. Use this microscope for the study and counting of large plankters such as mature microcrustacea. Include 10 × to 15 × paired occulars in combination with 1 × to 8 × objectives. This combination of optics bridges the gap between the hand lens and the compound microscope and provides magnification ranging from 10 × to 120 ×. Alternatively, use a good-quality zoom-type instrument with comparable magnification.

Inverted Microscope

The *inverted, compound microscope* often is used routinely for plankton counting in many laboratories. This instrument is unique in that the objectives are below a movable stage and the illumination comes from above, thus permitting viewing of organisms that have settled to the bottom of a chamber. Place samples in a cylindrical settling chamber having a thin, clear glass bottom. Chambers of various capacities are available ; the appropriate size depends on the density of organisms. After a suitable period of settling count organisms in the settling chamber.

The major advantage of the inverted microscope is that by a simple rotation of the nosepiece a specimen can be examined (or counted) directly in the settling chamber at any desired magnification. Although not recommended, oil immersion objectives have some useful applications. No preparation or manipulation other than settling is required. Generally, examine a preserved sample. Techniques are available for samples with an abundance of organisms that tend to float.

Fluorescence Microscope

An *fluorescence microscope* may be either standard or inverted. It uses incident light to excite electrons in intracellular compounds, such as pigments or absorbed stains, with the energy emitted during electron return to the ground state being measured as fluorescent light. The technique has been applied to the microscopic identification of chlorophyll-containing cells (autotrophs) and non-pigmented heterotrophic plankton fluorescent stains such as primulin or proflavin also have been used to differentiate nannoplanktonic primary and secondary producers. Excitation and emission wavelengths are unique for each pigment and stain and require distinct light, filter combinations and light sources. Select the filter combinations for the particular application. Epifluorescence microscopy is particularly useful for the enumeration of picoplankton and heterotrophic flagellate populations common to most aquatic systems. Concentrate samples by membrane filtration. Use epifluorescence microscopy as a complementary procedure to standard light microscope counting techniques.

Microscope Calibration

Microscope calibration is essential. The usual equipment for calibration is a Whipple grid (ocular micrometer, reticle, or reticule) placed in an eyepicce of the microscope and a stage micrometer that has a standardized, accurately ruled scale on a glass slide. The Whipple disk has an accurately ruled grid sub-divided into 100 squares. One square near the center is subdivided further into 25 smaller squares. The outer dimensions of the grid are such that with a 10 × objective and a 10 × ocular, it delimits an area of approximately 1 mm^2 on the microscope stage. Because this area may differ from one microscope to another, carefully calibrate the Whipple grid for each microscope.

With the ocular and stage micrometers parallel and in part superimposed, match the line at the left edge of the Whipple grid with the zero mark on the stage micrometer scale. Determine the width of the Whipple grid image to the nearest (0.01 mm from the stage micrometer scale. Should the width of the image of the Whipple grid be exactly 1 mm (1000 μm), the larger squares will be 1/10 mm (100 /μm) on a side and each of the smaller squares 1/50 mm (20 μm).

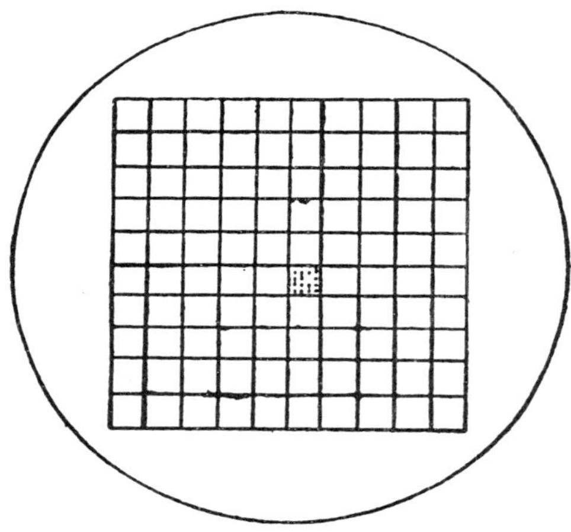

Figure 4.3 : ***Ocular micrometer ruling. A Whipple micrometer reticule is illustrated.***

When the microscope is calibrated at higher magnifications, the entire scale on the stage micrometer will not be seen, make measurements to the nearest 0.001 mm. Additional details for calibration are available.

MICRO COUNTING TECHNIQUES

Counting of Phytoplankton

Counting units

Some phytoplankton are unicellular while others are multicellular (colonial). The variety of configurations poses a problem in enumeration. For example, should a four-celled colony of *Scenedesmus* be reported as one colony of four individual cells? Listed below are suggestions for reporting:

Enumeration Method	*Counting Unit*	*Reporting Unit*
Total cell count	one cell	Cells/ml.
Natural unit count (clump count)	One organism (any unicellular organism or natural colony)	Units/ml.
Areal standard unit count	400 μm^2	Units/ml.

Making a total cell count is time-consuming and tedious, especially when colonies consist of thousands of individual cells. The natural unit or clump is the most easily used system; however, it is not necessarily accurate because sample handling and preserving may dislodge cells from the colony. The unit method also may not be quantitatively accurate nor reflect abundance of biomass or biovolume. Whatever method is chosen, identify it in reporting results.

If the distribution of organisms is random and the population fits a Poisson distribution, the counting error may be estimated. For example, the approximate 95% confidence limits, as a percentage of the number of units counted (N), equals:

$$\frac{2}{\sqrt{N}}(100\%)$$

Thus, if 100 units are counted, the 95% confidence limits approximate-20%. For a count of 400 units, the limits are about 10%.

Counting procedures

To enumerate plankton use a counting cell or chamber that limits the volume and area for ready calculation of population densities.

When counting with a Whipple grid, establish a convention for tallying organisms lying on an outer boundary line. For example, in counting a "field" (entrire Whipple square), desinate the top and left boundaries as "no-count" sides, and the bottom and right boundaries count" sides. Thus, tally every plankter technique a "count" side from the inside or outside but ignore any touching a "no-count" side. If significant numbers of filamentous or other large forms cross two or more boundaries of the grid, count them separately at a lower magnification and include their number in the total count.

Do not count dead cells or broken bottom frustules. Tally empty centric and pennate diatoms separately as "dead centric diatoms" or "dead pennate diatoms" to use in converting the diatom species proportional count to a count per milliliter.

Magnification is important in phytoplankton identification and enumeration. Although magnifications of 100 × to 200 × are useful for counting large organisms or colonies, much higher magnifications often are required. It is useful to categorize techniques for phytoplankton counting according to the magnifications provided.

Low-magnification (up to 200 ×) methods

The SedgwickRafter (S.R) cell is a device commonly used for plankton counting because it is easily manipulated and provides reasonable reproductible data when used with a calibrated microscope equipped with an eyepiece measuring device such as the Whipple grid.

The greatest disadvantage associated with the cell is that objectives providing high magnification cannot be used As a result, the S-R cell is not appropriate for examining nannoplankton. The S-R cell is approximately 50 mm long by 20 mm wide by 1 mm deep. The total area of the bottom is approximately 1000 mm^2 and the total volume is approximately 1000 mm^3 or 1 mL. Carefully check the exact length and depth of the cell with a micrometer and calipers before use.

(1) *Filling* cell-Before Filling the S-R cell with sample, place the cover glass diagonally across the cell and transfer sample with a large-bore pipet. Placing cover slip in this manner will help prevent formation of air bubbles in cell corners. The cover slip often

will rotate slowly and cover the innner portion of the S-R cell during filling. Do not overfill because this would yield a depth greater than 1 mm and produce an invalid count. Do not permit large air spaces caused by evaporation to develop in the chamber during a lengthy examination. To prevent formation of air spaces, occasionally place a small drop of distilled water on edge of cover glass.

Before counting let the S-R cell stand for at least 15 min to settle plankton. Count *plankton* on the bottom of the S-R cell. Some phytoplankton, notably some blue-green algae or motile flagellates in unpreserved samples, may not settle but rise to the underside of the cover slip. When this occurs, count these organisms and add to total of those counted on the cell bottom to derive total number of organisms. Count algae in strips or fields.

(2) *Strip counting—A* "strip" the length of the cell constitutes a volume approximately 50 mm long, 1 mm deep, and the width of the total Whipple grid.

The number of strips to be counted is a function of the precision desired and the number. of units (cells, colonies, or filaments) per strip. Drive number of plankton in the S-R cell from the following

$$\text{No./mL} = \frac{\text{C} \times 1000 \text{ mm}^3}{\text{L} \times \text{D} \times \text{W} \times \text{S}}$$

where

C = number of organisms counted.

L = length of each strip (S-R cell length), mm.

D = depth of a strip (S-R cell depth), ntm.

W = width of a strip (Whipple grid image width), mm. and

S = number of strips counted.

Multiply or divide number of cells per milliliter by a correction factor to adjust for sample dilution of concentration.

(3) *Field counting* — On samples containing many plankton (10 or more plankters per field), make field counts rather than strip counts. Count plankters in random fields each consisting of one

Whipple grid. The number of fields counted will depend on plankton density and statistical accuracy desired. Calculate the number of plankton per milliliter as follows:

$$\text{No./mL} = \frac{C \times 1000\ \text{mm}^3}{A \times D \times F}$$

where

C = number of organisms counted.

A = area of a field (Whipple grid image area), mm.

D = depth of a field (S-R cell depth), mm. and

F = number of fields counted.

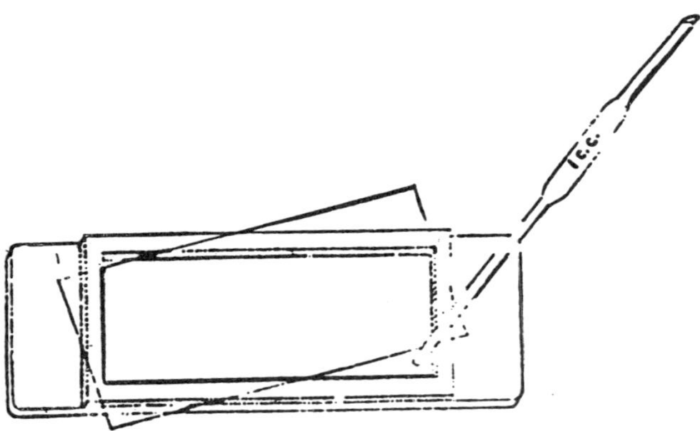

Figure 4.4 : ***Counting cell (Sedgwick-Rafter), showing method of filling.***

Multiply or divide the number of cells per milliliter by a correction factor to adjust for sample dilution or concentration.

Intermediate magnification (low to 500 ×) methods

The Palmer-Maloney (P-M) nannoplankton cell[3] is designed specifically for nannoplankton enumeration. It has a circular chamber with a 17.9-mm diam, 0.4-mm depth, and 0.1-mL volume. The shallow depth permits use of 40 to 45 × objectives with sufficient working distance. The principal disadvantage of the P-M cell is that these magnifications (400 to 450 ×) often are insufficient for nannoplankton identification and enumeration.

Because a relatively small sample portion is examined in the P-M cell do not use it unless the sample contains a dense population (10 or more plankters per field). Such a small sample portion from a less dense population causes serious under-estimation of density.

Introduce sample with a pipet into one of the 2- by 5-mm channels on the side of the chamber with the cover slip in place. After a 10-min settling period count the plankters in random fields, with the number of fields depending on density and variety of plankton and the statistical accuracy desired. Strips may be counted in this or any other circular cell by measuring the effective diameter and counting two perpendicular strips that cross at the center. Calculate the number per milliliter as follows

$$\text{No./mL} = \frac{C \times 1000\ \text{mm}^3}{A \times D \times F}$$

where

C = number of organisms counted.

A = area of a field (Whipple grid image) mm^2.

D = depth of a field (P-M cell depth), mm. and

F = number of fields counted.

Multiply or divide the number of cells per milliliter by a correction factor to adjust for sample dilution or concentration.

Another readily available chamber is the standard medical hemacytometer used for enumerating blood cells. It has a ruled grid machined into a counting plate and is fitted with a ground-glass cover slip. The grid is divided into 1-mm^2 divisions; the chamber is 0.1 mm deep. Introduce sample by pipet and view under 450 × magnification. Count all cells within the grid. The chamber comes from the manufacturer with a detailed instruction sheet containing directions on calculations and proper usage. A disadvantage to these counting cells is that the sample must have a very high plankton density to yield statistically reliable data.

High-magnification methods

Examination of phytoplankton at high magnification requires

the use of oil immersion objectives. Suitable procedures include using inverted microscope chambers, membrane filter mounts, sedimented slide mounts, the Lackey drop method, and diatom mounts.

(1) *Inverted microscope counts—Prepare a* sample for examination by filling the settling chamber. After the desired settling transfer the chamber to the microscope stage. Count perpendicular strips across the center of the bottom cover glass. Strip counts may be made by using a Whipple grid or special counting oculars that have a pair of adjustable parallel hairs and a single cross hair. Determine the width of the strip with a stage micrometer and tally organisms as they pass the single cross hair that functions as a reference point. Hold strip width constant fo any series of samples. Alternatively examine random nonoverlapping fields until at least 100 units of the dominant species are counted. For highest accuracy, particularly because algae distribution may be nonuniform, count the entire chamber floor. Alternatively, make a random fieldminimum count to attain a precision level of at least 85 %.

$$\text{Strip count (No./mL)} = \frac{C \times A_t}{L \times W \times S \times V}$$

where

C = number of organisms counted.

A_t = total area of bottom of settling chamber, mm^2,

L = length of a strip, mm.

W = width of a strip (Whipple grid image width), mm.

S = number of strips counted, and

V = volume of sample settled, mL.

$$\text{Field count (No./mL)} = \frac{C \times A_t}{A_f \times F \times V}$$

where

A_f = area of a field (Whipple grid image area), mm^2.

F = number of fields counted.

and other terms are as defined above.

(2) *Membrane filter mounts* -- Concentrate sample and prepare membrane filter.

Examine samples, concentrated on unlined membrane filters and mounted in oil as described above. Count enough random fields to ensure desired level of statistical accuracy. Select magnification level and size of microscope field (quadrat) such that the most abundant species appear in at least 70% but not more than 90% of microscopic fields examined (80% is optimum). Adjust microscope field size by using part or all of the Whipple grid. Examine 30 random microscope fields and record number of fields in which each species occurred. Report results as organisms per mililiter, calculated as follows:

$$\text{No./mL} = \frac{N \times Q}{V \times D}$$

where :

N = density (organisms/field) from Table follows:

Q = number of fields per filter.

V = milliliters filtered, and

D = dilution factor (0.96 for 4% formation preservative).

CONVERSION TABLE FOR MEMBRANE FILTER TECHNIQUE (BASED ON 30 SCORED FIELDS)

Total Occurrence	*F** %	*N*[+]
1	3.3	0.03
2	6.7	0.07
3	10.0	0.10
4	13.3	0.14
5	16.7	0.18
6	20.0	0.22
7	23.3	0.26

8	26.7	0.31
9	30.0	0.35
10	33.3	0.40
11	36.7	0.45
12	40.0	0.51
13	43.3	0.57
14	46.7	063
15	50.0	0.69
16	53.3	0.76
17	56.7	0.83
18	60.0	0.91
19	63.3	1.00
20	66.7	1.10
21	70.0	1.20
22	73.3	1.32
23	76.7	1.47
24	80.0	1.61
25	83.3	1.79
26	86.7	2.02
27	90.0	2.30
28	93.3	2.71
29	96.7	3.42
30	100.0	?

$$*F = \frac{\text{total number of species occurence} \times 100}{\text{total number of fields examined}}$$

+N = number of organisms per field

(3) *Diatom mounts*—Prepare samples

For diatom species proportional count, examine diatom samples under oil immersion at a magnification of at least 900 ×. Scan lateral strips the width of the Whipple grid until at least 250 cells are counted. Available time and accuracy required dictate the number of cells to be counted. Determine percentage abundance of each species from tallied counts and calculate counts per milliliter of each species by multiplying per cent abundance by total live and dead diatom count obtained from the plankton counting chamber. For greater accuracy distinguish besween living and dead diatoms at the species level.

(4) *Phytoplankton staining technique*—Staining algae permits differentiation between "live" and "dead" diatoms. This permits enumerating total phytoplankton in a single sample without sacrificing detailed diatom taxonomy. It also results in permanent reference slides. The procedure is most useful when diatoms are major components of phytoplankton and it is important to distinguish between living and dead diatoms.

Preferably perserve samples in Lugol's solution or alternatively in formalin. For analysis thoroughly mix the sample and filter a portion through a 47-mm-diam membrane filter (pore diam 0.45 or 0.65 μm). Use a vacuum of 16 to 20 kPa and never let sample dry. Add 2 to 5 mL aqueous acid fuchsin solution (dissolve 1 g acid fuchsin in 100 mL distilled water to which 2 mL glacial acetic acid has been added ; filter) to the filter and let stand for 20 min. After staining, filter sample, wash briefly with distilled water, and filter again. Administer successive rinses of 50%, 90%, and 100% propanol to the sample while filtering. Soak for 2 min in a second 100% propanol wash, filter, and add xylene. At least two washes are required; let the final one soak 10 min before filtering. Trim the xylene-soaked filter and place on a microscope slide on which there are several drops of mounting medium. Apply several more drops of medium to top of filter and install a cover glass. Carefully squeeze out excess mounting medium. Make the final mount permanent by lacquering the edges of the cover glass.

Count organisms using the most appropriate magnification. "Live" diatoms typically are red while "dead" ones are unstained. Oil immersion is necessary for species identifications of diatoms and many other algae. Count either strips or random fields and calculate plankton densities per milliliter:

$$\text{No./mL} = \frac{C \times A_t}{A_c \times V}$$

where

C = number of organisms counted.

A_t = total area of effective filter before trimming and mounting.

A_c = area counted (strips or fields), and

V = volume of sample filtered, mL.

COUNTING OF ZOOPLANKTON

Subsampling

Count entire samples with <200 zooplankters without subsampling. Most zooplankton samples will contain more organisms than can be enumerated practically, therefore, use a subsampling procedure. Before subsampling, remove and enumerate all large uncommon organisms such as fish larvae in fresh water or coelenterates, decapods, fish larvae, etc., in salt water.. Subsample by the pipet or splitting method.

Alternatively, subsample by splitting with any of a number of devices of which the Folsom plankton splitter is best known. Level splitter before using. Place sample in the splitter and divide into subsplits. Rinse splitter into the subsamples. Repeat until a workable number (200 to 500 individuals) is obtained in a subsample. Exercise care to provide unbiased splits. Even when using the Folsom splitter unbiased subsamples cannot be unquestioningly assumed ; therefore, count animals in several subsamples from the same sample to verify that the splitter is unbiased and to determine the sampling error introduced by using it.

Another method permits abundance estimates of more equivalent levels of precision among taxa than obtained with either the Hensen-Stempel pipet of the Folsom splitter. Normal counting procedures tally organisms on the basis of their abundance in a sample. Therefore, in a sample with a dominant organism making up 50% of total numbers, the tally of the dominant taxon will be large and have a small error. However, error about the subdominants will increase as the tally of each taxon decreases, By accepting one level of precision, the technique has been developed to obtain the same error about dominants and subdominants, permitting quantitative comparisons between taxa over successive times or between stations.

Enumeration

Using a compound microscope and a magnification of 100 ×, enumerate small zooplankton (protozoa, rotifers, and nauplii) in a 1- to 5-mL clear acrylic plastic counting cell fitted with a glass cover slip. For larger, mature microcrustacea use a counting chamber holding 5 to 10 mL. A Sedgwick-Rafter cell is not suitable because of size. An open counting chamber 80 by 50 mm and 2 mm deep is desirable ; however, an open chamber is difficult to move without jarring and disrupting the count. A mild detergent solution placed on the chamber before counting reduces organism movements or special counting trays with parallel or circular grooves or partitions can be used. Count microcrustacea with a binocular dissecting micro. scope at 20 × to 40 × magnification. If identification is questionable, remove organisms with a microbiological transfer loop and examine at a higher magnification under a compound microscope.

Report smaller zooplankton as number per liter and larger forms as number per cubic meter:

$$\text{No./m}^3 = \frac{C \times V'}{V'' \times V'''}$$

where

C = number of organisms counted,

V′ = volume of the concentrated sample, ml,

V″ = volume counted ; mL. and

V‴ = volume of the grab sample, m^3,

To obtain organisms per liter divide by 1000.

(2) Quantitative estimation of plankton density

(BY TRANSACT METHOD)

Apparatus:

(*i*) Micrometers

(*ii*) Microscope

(*iii*) Slides

(*iv*) Cover glass (22 × 22 mm or 18 × 18 mm).

Procedure:

(1) Determination area of one vision field:

First setup the oculometer into eye piece and measure, bow many divisions of oculometer cover the total area. The value of one division of oculometer is known (1 division of oculometer=.015 mm), so total area can be determined by calculation.

Total No. of divisions of oculometer × .015.

The area of vision field is not in square, it is circular so by using following formula the total area of one vision fieldca be determined:

Total area of vision field = πr 2(in sq. mm)

Where, $\pi = 3.14$; r = Radius.

$$r = \frac{A}{2}$$

where A = value of divisions of oculometer (in mm) covered by one vision field.

(II) Determination of total number of vision fields in one Transact:

(*i*) Take any material on a slide for marking and put cover glass (22 × 22 or 18 × 18 mm) on the material which taken on slide.

(*ii*) Placed the slide under microscope and start the counting of number of vision fields from one corner to the other corner of cover glass. By this way take 3 or 4 measurements and calculate mean value of number of vision fields for one Transact.

(III) Determined of total area of one Transact;

Total area of one vision field is known (from step I) and the total number of vision fields in one Transact are also known (from step II), so total area of one Transact can be determined by following calculation.

Total area of one = X × Y (in mm^2)

Where,

X = Total No. of vision fields in one Transact.

Y = Total area of one vision field.

(IV) Estimation of No. of Planktons:

(*i*) Take 40 liters of water sample and filter it through plankton net.

(*ii*) Collect the sample from tube of plankton net and make up it to 50 ml. (So this 50 ml of sample represents the 40 litres of water supply).

(*iii*) Take 1 ml (about 2 drops) of this sample on slide and cover wtih cover glass (22 × 22 or 18 × 18 mm.)

(*iv*) Placed the slide under microscope and count the planktons for 5 or 6 different trnasacts and calculate the mean vlaue (No. of planktons) for one Transact.

(*v*) Now the total area of one Transact (Step III), total no. of planktons in one Transact (step IV) and total area of cover glass (22 × 22 mm or 18 × 18 mm) are known, so total number of planktons under cover glass (as per ml because 1 ml of material was taken on slide) can be determined by the following calculation:

$$\text{No. of planktons (per ml)} = \frac{\text{Area of cover glass}}{\text{Area of one transact}} \times a.$$

Where,

Area of cover glass = (22 × 22 or 18 × 18 mm)

Area of one Transact = From Step III

a = Mean value of No. of planktons for one Transact.

(*vi*) By above calculation, the number of planktons per ml can be determined and with the is value the no. of planktons for 50 ml sample (which represents 40 litres of sample) can be determined by following calculation.

No. of planktons in 40 litres sample or 50 ml = No. of planktons per ml × 50.

(*vii*) By above calculation, the number of planktons in 0 litres can be determined and with this value the number of planktons per litre can be determined.

$$\text{No. of planktons per litre} = \frac{\text{No. of planktons in 40 litres}}{40}$$

Result:

Planktons per liter.

ESTIMATION OF BIOMASS

Biomass is the weight of all living materials in a unit area at an given time in a particular aquatic body. It is done in the following ways :

(*i*) Wet weight

(*ii*) Dry weight

(*iii*) Ash weight

(i) By Wet Weight:

In this method the weight of wet biomass is determined

Procedure:

(*i*) Take initial wt. of wet filter paper.

(*ii*) Now filter 1 liter of water sample through filter paper.

(*iii*) Take the weight of filter paper.

Calculation:

Wet wt. of biomass=A – B (mg/liter) where,

A=Total wt. of filter paper with biomass.

B=Initial wt. of wet filter paper.

(ii) By Dry Weight

In this method weight of dry biomass is determined.

Procedure:

(*i*) Take initial wt. of filter paper (dry).

(*ii*) Now filter 1 liter of water sample through filter paper.

(*iii*) Keep the filter paper in oven foi about 24 hrs. at 0°C temperature.

(*iv*) Taken out the filter paper from oven after 24 hrs. and taken the total weight of filter paper.

Calculation:

Total dry wt. of biomass—X—Y (mg/liter) where,

X = Total wt. of filter paper.

Y = Initial wt. of dry filter paper.

(iii) *By ash weight*

In this method wt. of ash of biomass is determined.

Procedure:

(*i*) Filter one liter of water sample through filter paper (Filter paper should be ashless)

(*ii*) Keep the filter paper in crucible and dry it.

(*iii*) Take the weight of dry filter paper with crucible (1w).

(*iv*) Now keep the crucible with filter paper in muffle furnace for hrs. at 00 to 900°C temperature.

(*v*) After hrs. take out the crucible from muffle furnace and take the weight (Fw).

(*vi*) Substract the weight (Fw) of crucible from the initial weight (1w) of crucible. This will give the weight of biomass in the form of organic matter.

Calculation:

Wt. of Biomass—Iw—Fw

(in the form of organic matter)

where,

Iw = Initial wt. of crucible with filter paper and biomass.

Fw = Final wt. of crucible with ash.

ESTIMATION OF PRIMARY PRODUCTIVITY

Light and Dark Bottle Method

Primary productivity:

The primary productivity of an ecosystem is defined as the rate at which radient energy is stored by photosynthesis of a producer organism into the organic substances, which are used as food material.

Gross Primary productivity (GEP):

This is the total rate photosynthesis including organic material last by respiration during the period. GEP=Total new organic matter + Respiratory loss.

Net Primary productivity (NEP)

This is rate of storage of organic matter in the plant in excess of that last by respiration during the period. (NEP) = New organic matter-Respiratory loss.

Principle:

Plants produce new organic carbon by the process of photosynthesis. In this process the oxygen is released by plants. So by estimating the oxygen which is released by plants in photosynthesis, the primary productivity can be estimated.

In this method light and dark bottles are used. For estimation of primary productivity, first the initial dissolved oxygen is determined, then (after a definite time) the dissolved oxygen of light and dark bottle is determined. In light bottle the level of dissolved oxygen (L) increases than initial level (I), because in this bottle photosynthesis takes place. The increasing value of dissolved oxygen in light bottle can be determined by substracting initial value of dissolved oxygen. This value is multiplied by conversion factor for carbon (.375). So this gives not primary productivity (NPP) because the value of respiratory loss (R) can be determined by estimating the dissolved oxygen (D) in the dark bottle. In dark bottle the value of dissolved oxygen decreases (because in this bottle photosynthesis does not take place, so the O_2 is not released but the respirations take place). The value of dissolved oxygen (D) in dark bottle is substracted from initial value of dissolved oxygen (I). It gives value of (I) respiration (R). If this value (R) is added to value of net primary productivity, it gives the value of gross primary productivity (GPP).

Apparatus:

(*i*) Light and dark bottles.

(*ii*) Rope or other means to suspend pairs of bottles in lake.

(*iii*) Other water analysis apparatus described under dissolved oxygen.

Reagents:

Same as for dissolved oxygen.

Procedure:

(*i*) Select station for experiment.

(*ii*) Take water sample in light and dark bottles and suspend the bottles in water body by rope or other means.

(*iii*) Note the initial time.

(*iv*) Take one more water sample at the same time in which estimate initial dissolved oxygen (I) by procedure given in estimation of dissolved oxygen.

(*v*) After three hours, take out the bottles from water body and estimate dissolved oxygen in water samples of both bottles.

Calculations:

(i) *Determination of Net Photosynthesis (NP):*

Net photosynthesis (NP) = L—I—mg/liter/3 hrs. (In form of mg of O_2/ 1).

where,

L = Value of dissolved oxygen in light bottle.

I = Initial value of dissolved oxygen.

(ii) *Determination of Respiration (R):*

Respiration (R)=I-D-mg/liter/3 hrs. (In form of mg of O_2/I).

where,

I = Initial value of dissolved oxygen.

D = Value of dissolved oxygen in dark bottle.

(iii) *Determination of Gross Photosynthesis (GP):*

Gross photosynthesis (GP) = NP+R—mg/l/3 hrs. (In form of mg of O_2/1).

where,

NP = Value of Net photosynthesis.

R = Respiration.

(iv) Determination of Primary Productivity

Primary productivity can be determined by multiplying the rate of photosynthesis by conversion factor for carbon (.375).

Net primary productivity= NP × .375—gmC/m^3/3 hrs. **(N PP).**

Gross Primary productivity= GP × .375-gm C/m^3/3 hrs. (GPP).

where,

NP = Net photosynthesis. GP = Gross photosynthesis.

C^{14} method:

This method is used for accurate estimation of PP. and facilitates estimations of Research level. Only the principle is given here. Two glass bottles (a light and a Dark bottle) are used in micro winkler bottles (About 25 ml) and water Samples are taken. A suitable amount of C^{14} is injected or inoculated into these bottles. After the experimental period the uptake of C^{14} is estimated in a Scintillation Counter by which the amount of new organic carbon can be estimated.

ESTIMATION OF CHLOROPHYLL A

The following is a general survey method for estimating the concentration of phytoplankton chlorophyll a pigment in a volume of water. The value of such data is uncertain, since many variables are present.

Apparatus:

(*i*) Spectrophotometer.

(*ii*) Rubber stoppers or corks

(*iii*) Test tubes

(*iv*) Membrance filter unit with 0.8 ,am, 47-mm filters.

(*v*) Centrifuge with stoppered glass centrifuge tubes.

Reagents:

90% alkalized aceton. Using reagent grade acetone and distilled water, prepare a 90% solution. Add about one spoonful of powdered magnesium carbonate per liter of acetone solution.

Procedure:

(i) Take water samples at several random location over lake and at several depths down to limit of photiczone. Usually 1-liter samples will be adequate.

(*ii*) Filter portion of well-shaken sample through 0.8 $_{\mu}$m, 47-mm membrane filter.

(*iii*) Roll up or fold filter with plankton inside. Place filter in bottom of dry test tube.

(*iv*) Measure volume of water filtered. Suve remainder of unfiltered sample.

(*v*) Add 20 ml of 90% alkalized acetone to sample tube containing filter and plankton, Stopper tube with an acetone-leached stopper. Shake until filter is dissolved.

(*vi*) Allow extraction to proceed in a dark refrigerator. An overnight extraction is preferable.

(*vii*) If any turbidity is apparent in extract, Centrifuge extract in stoppered tubes.

(*viii*) Use a blank of 20 ml of acetone containing the same number of filters used for sample, and measure absorbance of sample and blank at 663 nm. In acetone extracts the absorbance of this turbidity blank may be appreciable.

Calculation:

For grass green extracts:

Chlorophyll A concentration.

(mg per liter of extract) = absorbance × 7.5*

(*7.5=is used for 1-inch cuvettes; 13.4 is used for 1 cm cuvettes).

Chlorophyll A is the chief chlorophyll measured at 63 nm; however, the band width of the B & L spectronic 20 probably also measures some chlorophyll B when it is present. Chlorophyll a concentration of original lake water (mg/m^2)

$$= \frac{\text{Chlorophyll in extract (mg / liter)} \times \text{extract volume (ml)}}{\text{filtered volume (liters)}}$$

5

Biological Water Quality Monitoring

Biological water quality monitoring is the best method to know the water quality without causing any changes in it. This method has now become the internationally recognised and most widely used. This can be done by the following method;

DIVERSITY INDICES

The diversity indices are calculated from the abundance data of the organisms and serve as a very good indicator of pollution. Some of the common diversity indices are:

Kothe's Species Deficit Index

This index is based on the principle that in a flowing ecosystem the number of species decreases after they are exposed to some Pollutant discharge. In this method the number of species of either a particular group (e.g. algal or macroinvertebrates) or of all the group are counted at the polluted and non-polluted points and index is calculated by using following formula

$$\text{Kothe's species deficit} = \frac{A_1 - A_x}{A_1} \times 100$$

Where A_1 = No. of species at the unpolluted site

A_x = No. of species at the polluted site, downstream.

It gives the data in a percentage linear scale and is very useful in indicating the consequences of point sources of waste water discharges. It requires practically no equipments. The higher the value of deficit more is the level of pollution at that site.

General Species Index

It is an excellent index to determine the level of pollution in both flowing and standing water bodies (Odum 1960). It can be calculated by using following formula:

$$O.I. = \frac{\text{Total no. of species encountered in the sample}}{\text{Total no. of individuals of all the species}} \times 1000$$

Here again either a particular group can be taken alone or all the groups are taken together.

Shannon's diversity Index

$$H = -\sum \frac{n_1}{n} \text{Ln} = \frac{n}{n}$$

Where H = Diversity index in bits/individual

n_1 = Number of individuals in species of a population or community (individual density of one Sp.)

n = Number of individuals in a sample from a population (density of all the species)

Ln= Log normal

Although any biological group can be used to calculate the diversity index, it is better to use macroinvertebrates which stay in water and are not carried to long distances in flowing waters. However, in stagnent ecosystems, this index can very well be used even with phytoplankton. The index is likely to be influenced by sampling method, sample size, depth of sampling, duration of sampling, time of the year and taxonomical knowledge. So, adequaute care has to be taken of these factors before deriving a conclusion.

BIOLOGICAL INDICES

Unlike diversity indices, the biotic indices are dependent on the indicator species rather than on the community structure. They have been extensively used in water pollution monitoring programmes, but Washington feels that they are unlikely to yield universally applicable results due to the taxonomic variability in various countries. They are essentially dependent on 'bioindicator' or indicator species. According to Washington, an indicator species is the one which is sensitive to pollution (For example disappearance of this species from polluted areas can be seen). Some of the common biotic indices are described here.

Goodnight and Whitley's Index

$$\text{G \& W Index} = \frac{\text{No. of tubificideae recoreded}}{\text{Total number of benthic Organisms recorded.}} \times 100$$

Explanation:

Index value		Pollution level
80	:	Highly polluted
60 – 80	:	Doubtful (polluted)
Less than 60	:	Class areas.

King and Ball's Index

$$\text{K \& B Index} = \frac{\text{Insect wt.}}{\text{Tubified wt.}}$$

Not much taxonomic knowledge is required to calculate this index.

Beak's River Index

Beak, based on his long term observations on macroinvertabrate populations has derived this index. The Index is dependent on the presence, absence of abundance of sensitive or tolerant macroinvertebrate species and is calculated as per the following chart.

TABLE-5.1.

Pollution status	*Biotic Index*	*Type of Macroinvertebrates Community*	*Fishery potential*
Upolluted	6	Sensitive, facultative and tolerant predators, filter and detritus feeder all represented but no species well developed.	All normal fisheries

Slight to Moderate pollution	4-5	Sensitive predators and herbivores reduced in population density or absent. Facultative predators herbivores and possibly filter and detri-	Most sensi tive fish species reduced in number or missing. tus well
developed and		increasing in numbers as index decreases.	
Moderate pollution	3	All sensitive species absent and facultative predators **(Hirudinea)** absent or scarce. Predators of	Only coarse fishery maintained family
Pelopinae and		herbivores of Tendipedidae present in fairly large population densities.	
Moderate to Heavy pollu- reduced in num- tion	2	Facultative and tolerant pollution bers if, pollution is toxic, if the pollution is of organic kind, few species which can thrive in low oxygen are present.	Only higly species tolerant fish species are present
Heavy pollution	1	Only most tolerant detritus feeders (Tubificidae) present in large numbers,	Very little if any fishery
Severe pollution	0	No macroinvertebrates present.	No fish

In calculating this index, whole macroinvertebrate fauna is used.

Trent's Biotic Index

This index was developed by Woodiwise when he was working on Trent River. Woodiwise devised a scheme in which the number of groups (of defined taxa) of benthic macroinvertebrates was related to the presence of six key organisms found in the fauna. These organisms are plecopteran, nymphs, ephemeropteran, mympho, trichopteran, larvae, Gammarus, Asellus and tubited/red chironomus larvae. Following is the detailed procedure of calculating this index.

The Trent Biotic Index

TABLE-5.2 : CLEAN

Organisms in order of tendency to disappear as degree of pollution increases		*Total number of groups present*				
		Column. 1	*2*	*3*	*4*	*5*
Row		0-1	2-5	6-10	11-15	16+
		Biotic index				
1. Plecoptera larvae	More than one species	—	7	8	9	10
2. Present	One species only	—	6	7	8	9
3. Ephemeroptera larvae	More than one species	—	6	7	8	9
4. Present	One species only	—	5	6	7	8
5. Trichoptera larvae	More than one	—	5	6	7	8
6.	One species only	4	4	5	6	7
7. Gammarus present	All above					
8.	Species absent	3	4	5	6	7
9. Asellus	All above species absent	2	3	4	5	—
10. Tubificid worn and/or red Chironomid larvae present	All above species absent	1	2	3	4	—

11. All above types absent	Some Organisms such as Eristalis tenex not requiring dissolved oxygen may be present.	0	1	2	—	—

Polluted

**Beatis rhodani* excluded

+*Beatis rhodani* (Ephem.) is counted in this section for the purpore of classification.

The term 'Group' used for purpose of the biotic index means any one of the species included in the following list of organisms or sets of organisms:

Each known species of plotyhelminthes (flatworms)

Annelida (worms excluding genus Nais)

Genus Nais (worms)

Each known species of Hirudinae (leeches)

Each known species of Mollusca (snails)

Each known species of Crustacea (hog louse. shrimps)

Each known species of Plecopotera (stone fly)

Each known genus of Epheme-roptera (may-fly, excluding Baetis rhodani)

Baetis rhodani (may-fly)

Each family of Trichoptera (Caddis-fly)

Each species of Megaloptera larvae (alder-fly)

Family Chirhnomidae (midge larvae except Chiron-omus riparius)

Chironomus riparius (blood worms)

Family Simulidae (black fly larvae)

Each known species of other fly larvae

Each known species of Coleoptera (Beetles and bettle larvae) Each known species of Hydracarina (Water mites)

The biotic index ranges from 10 for clean water to 0 for polluted water.

Example:

First of all the number of groups present in the sample are identified.

For a given sample

Platytheminthis (2) + Nais (1) + Hirudinae (1) + Crustacea (2) + mullusca (1) + Trichoptera (3) + Megaloptera (1) +Chironomus (1) + Other fly larvae (0) + Coleoptera (1) + Simulidae = 17.

Now there are 16 + groups present, so in the table we have to use the final column, (16 +). Since plecoptera larvae are not present in our hypothetical sample, we can proceed to row 3, but even ephemeroptera larvae are absent in our sample, so we proceed to row 5, since Trichoptera larvae are present and more than 1 species is present, the value of the index will become 8, which indicates quite high pollution.

Nygaard's Algal Indices

Nygaard proposed five indices to evaluate the organic pollution of a water body on the basis of algal groups (Myxophycean index, chlorophycean index, diatom index, eulglenophycean index and compound index). These indices have been developed on the basis of the fact that various algal groups have different tolerance to organic pollution and nutrient enrichment. As a general rule, the *Cyanophyta*, the centric diatoms (the diatoms with redical symmetry), and most *Chlorococcales* are commonly found in the eutrophic waters, while the pennate diatoms and the desmids (order *Zygenematales* of *Chlorophyta*), are commonly found in oligotrophic waters. For the calculation of these indices, all the algal genera present in a water sample are identified and put in the formulae provided in Table 5.3 the oligotrophic or eutrophic status of the sample can be ascertained as per the original values provided by Nygaard (Table 5.3). These indices are however, not applicable to the waters polluted by the effluents other than the organic or nutrient entriching kind in nature, e.g. toxic pollutants, pesticides, etc.

TABLE-5.3 : CALCULATION OF NYGAARD INDICES.

Index	Calculation	Oligotrophic	Eutrophic
Myxophycean	Myxophyceae (Cyanophyceae) / Desmideae	0.0-0.4	0.1-3.0
Chlorophycean	Chlorococcales / Desmideae	0.0-0.7	0.2-9.0
Diatom	Centric Diatoms / Pennate diatoms	0.0-0. 3	0.0-1.75
Euglenophycean	Euglenophyta / Myxophyceae + Chlorococcales	0.0-0.7	0.0-1.0
Compound	Myxophycease+ Chlorococcales+ Centric diatoms+ Eglenophyta / Desmideae	0.01-1.0	1.2-2.5

Palmer's Algal Pollution Indices

Palmer made the first major attempt to identify and prepare a list of genera and species of algae tolerant to organic pollution. He prepare a list of 60 genera and 80 species tolerant to organic pollution. Tables on page 102 provides the name of these algae in the order of decreasing tolerance.

Palmer's Algal Genus Index

For the calculation of this index, Table 5.4 is taken is use. This table provides algal genera most tolerant to organic pollution and a number is assigned to each of them depending on their relative tolerance. The algae present in a water sample are identified and the genera present from this list are noted (other genera ignored).

An algae is called present when 50 or more individuals of it are present in one ml of water. The numbers scored by each genera are totalled to get the value of algal genus index. A score of 20 or more for a sample is indication of organic pollution, while a score.

TABLE 5.4 : SIXTY MOST POLLUTION TOLERANT GENERA OF ALGAE, IN ORDER OF DECREASING EMPHASIS.

No.	*Genus*	*Group*	*No.*	*Genus*	*Group*
1.	Euglena	E	31.	Saurirella	D
2.	Oscillatoria	B	32.	Stephanodiscus	D
3.	Chlamydomonas	G	33.	Eudorina	G
4.	Scenedesmus	G	34.	Lyngbya	B
5.	Chlorella	G	35.	Oocystis	G
6.	Nitzschia	D	36.	Agmenellum	B
7.	Navicula	D	37.	Spirulina	B
8.	Stigeoclonium	G	38.	Pyrobotrys	X
9.	Synedra	D	39.	Cymbella	D
10.	Ankistrodemus	G	44.	Actinastrum	G
11.	Phacus	E	41.	Coelastrum	G
12.	Phormidium	B	42.	Cladophora	G
13.	Melosira	D	43.	Hantzschia	D
14.	Gomphonema	D	44.	Diatoma	D
15.	Cyclotella	D	45.	Spondylomorum	G
16.	Closterium	G	46.	Golenkinia	G
17.	Micractinium	G	47.	Achnanthes	D
18.	Pandorina	G	48.	Synura	CH
19.	Anacystis	B	49.	Pinnularia	D
20.	Lepocinclis	E	50.	Chlorococcum	G
21.	Spirogyra	G	51.	Asterionella	D

22.	Anabaena	B	52.	Cocconies	D
23.	Cryptomonas	CR	53.	Cosmarium	G
24.	Pediastrum	G	54.	Gonium	G
25.	Arthrospira	B	55.	Tribonema	G
26.	Trachelomonas	E	56.	Stauroneis	D
27.	Carteria	G	57.	Selenastrum	G
28.	Chlorogonium	G	58.	Dictyosphaerium	G
29.	Fragilaria	D	59.	Cymatopleura	D
30.	Ulothrix	G	60.	Crucigenia	G

E = Euglenophyta X = Xanthophyta

B = Blue-green algae (Cyanophyceae) D = Diatoms (Bacillariophyta)

C = Green algae (Chlorophyta) CH = Chrysophyta

CR = Cryptophyta

(Modification in group signs by authors)

TABLE 5.5: EIGHTY MOST POLLUTION TOLERANT SPECIES OF ALGAE IN THE ORDER OF DECREASING EMPHASIS.

1.	Euglena viridis	41.	Lepocinclis texta
2.	Nitzschia palea	42.	Euglena deses
3.	Oscillatoria limosa	43.	Spondylomorum quaternarium
4.	Scenedesmus quadricauda	44.	Phormidium uncinatum
5.	Oscillatoria tenuis	45.	Chamydomonas reinhardi
6.	Stigeoclomium tenue	46.	Chlorogonium euchlorum'
7.	Synedra ulna	47.	Euglena polymorpha
8.	Ankistrodesmus falcatus	48.	Phacus pleuronectus
9.	Pandorina morum	49.	Navicula viridural
10.	Oscillatoria chlorina	50.	Phormidium autumnale
11.	Chlorella velgaris	51.	Oscillatoria lauterborni

12.	Arthrospira Jenneri	52.	Anabaena constricta
13.	Melosira various	53.	Euglena pisciformis
14.	Cyclotella meneghiniana	54.	Actinastrum hantzschi
15.	Euglena gracilis	55.	Synedra acus
16.	Nizschia acicularis	56.	Chlorogonium elongatum
17.	Navicula cryptocephala	57.	Synura uvella
18.	Oscillatoria princeps	58.	Cocconeis placentula
19.	Oscillatoria putrida	59.	Nitzschia sigmoidia
20.	Gomphonema parvulum	60.	Coelastrum microporum
21.	Hantzschia amphioxys	61.	Achnanthes minutissima
22.	Oscillatoria chalybea	62.	Cymatopleura solea
23.	Stephenodiscus hantzschi	63.	Scendesmus dimorphtis
24.	Euglena oxyuris	64.	Fragillaria crotonensis
25.	Closterium acerosam	65.	Anacystis cyanea
26.	Scendesmus obliques	66.	Navicula cuspidata
27.	Chlorella pyrenoidosa	67.	Scendesmus acuminatus
28.	Cryptmonas erosa	68.	Euglena intermedia
29.	Eudorina elegens	69.	Pediastrum duplex
30.	Eugiena acus	70.	Closterium leibeinii
31.	Surirella ovata	71.	Oscillatoria brevis
32.	Lepocinclis ovum	72.	Traehelomonas volvocina
33.	Oscillatoria formosa	73.	Dictyosphaerium pulchella
34.	Oscillatoria splendida	74.	Fragilaria capucina
35.	Phacus pyrum	75.	Cladophora glomerata
36.	Micractinium pusillum	76.	Cryptomonas ovata
37.	Agmenellum quadriduplicatum	77.	Gonium pectorale
38.	Melosira gramulata	78.	Euglena proxima
39.	Pediastrum boryanum	79.	Pyrobotrys
40.	Diatoma vulgare	80.	Tetraedron muticum

Of 15 to 19 is taken as probable evidence of high organic pollution. Lower figures indicate that the organic pollution is not high or the sampling has not been representative.

Example. If an algal sample is having the genera, Chlorella, Oscillatoria, Sigeoclonium, Synedra, Nitzschia, Chlamydomonas, and Navicula, the score according to Table 5. 6 shall be 3 + 5 + 2 + 2 + 3 + 4 + 3 = 22. This confirms high organic pollution of this water sample.

Palmer's Algal Species Index

The calculation an interpretation of this index is same for the algal groups genus index but here the species and not the genera are taken into consideration.

TABLE 5.6 : POLLUTION INDEX OF ALGAL GENERA.

Genera	Pollution Index	Genera Index	Pollution
Anacystis (Microcystis)	1	Micractinum	1
Ankistrodesmus	2	Navicula	3
Chlamydomonas	4	Nitzschia	3
Chlorella	3	Oscillatoria	4
Closterium	1	Pendorina	1
Cyclotella	1	Phacus	2
Euglena	5	Phormediom	1
Gomphonema	1	Scenedesmus	4
Lepocinclis	1	Stigeoclonium	2
Melosira	1	Synedra	2

6

Toxicological Testing Procedures

GENERAL TOXICOLOGICAL METHODOLOGY

The particular sequence of manipulations to which a given specimen is subjected depends on the nature *of* the investigation and the particular substances which are likely to be present.

In the clinical toxicology setting the usual question is whether drugs(s) is (are) present, and if so which and in what concentration(s). Each laboratory must design a protocol based on : (1) resources available ; (2) the local epidemiology (what are the popular substances in that area). The protocol described below which has proven optimal for a particular population is indicative of the type of manipulations required.

Acidic and neutral drugs in serum or urine are absorbed (Fig. 6.1, 6.2) on to charcoal at acid pH 5.0-6.0 and eluted subsequently into an organic phase (a mixture of ether, chloroform, isopropyl alcohol, isoamyl alcohol). The organic phase eluate is concentrated by air stream at 40-50°C, reconstituted with methanol and subjected to gas-liquid chromatography (GLC) or GC and mass spectrography (MS) sequentially (see the discussion of GC-MS which follows). Basic drugs are double back-extracted prior to GC or GC-MS, a procedure which must be done on each serum and/ or plasma sample, Serum is adjusted to pH 9.0-9.5, then extracted with an organic phase (a mixture l-chlorobutane and hexane isoamyl alcohol). The organic layer is removed to another container and extracted with 0.5 *N* HCI, so that the basic substances go to the aqueous lower) layer. After discarding the organic (upper), layer, the residual aqueous layer is adjusted to basic pH

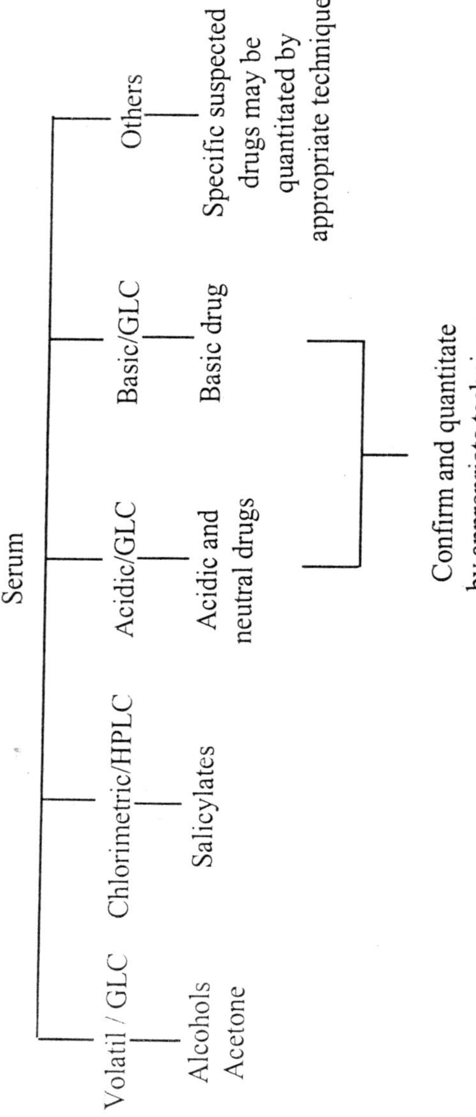

Figure 6.1: *Protocol for drug overlapse analysis of serum.*

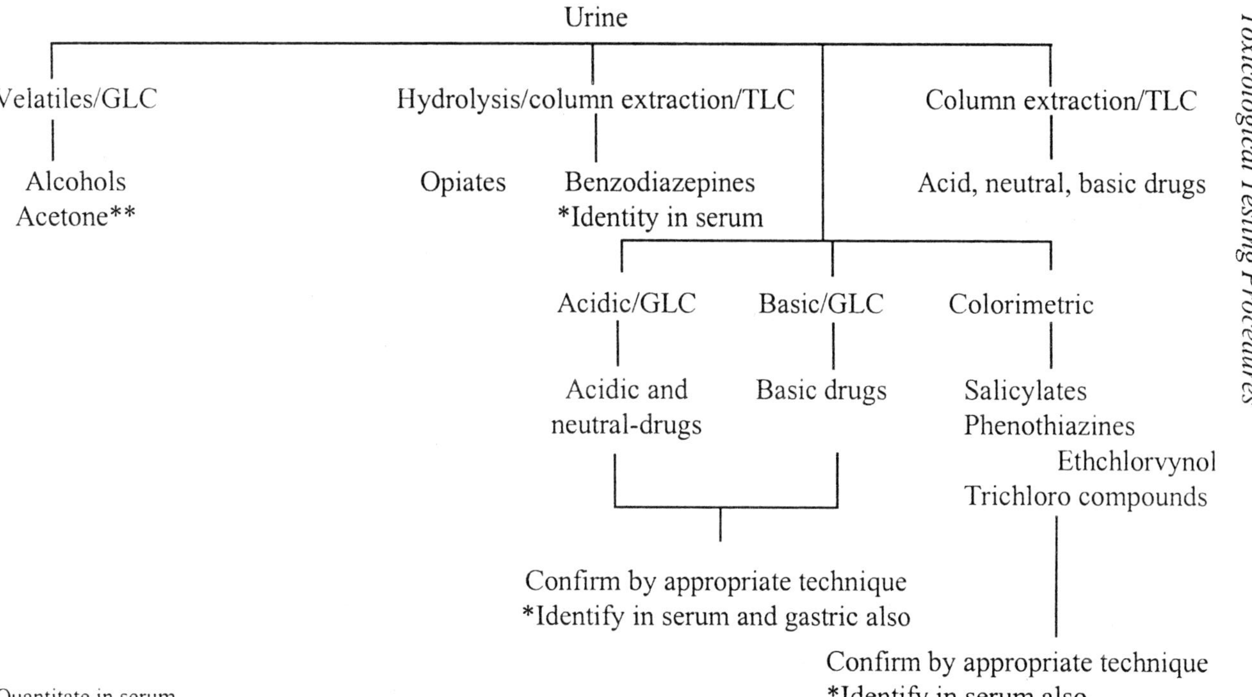

*Quantitate in serum
**Serum preferred

Figure 6.2 : *Protocol for drug overdose analysis of unine*

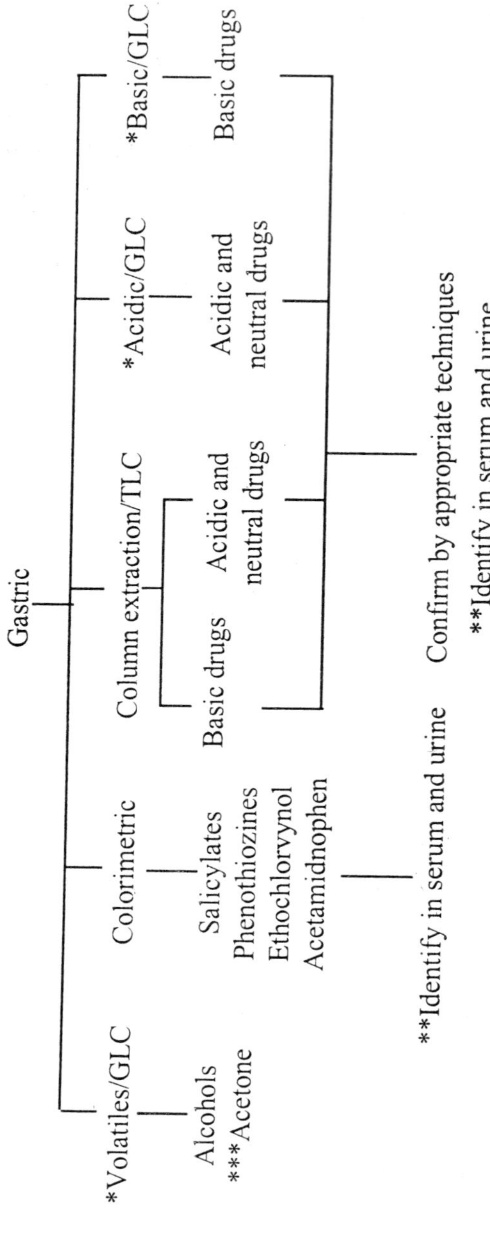

*Only if serum of urine is not provided
**Quantitates in serum
***Serum preferred

Figure 6.3 : *Protocol for drug overdose analysis of gastric contents.*

with 10 *N* sodium hydroxide and extracted again with the mixtured l-chlorobutane and hexane isoamyl alcohol. The organic layer is removed and concentrated by an air stream at 40-50°C after the addition of 0.5% HCl in methanol. It is subjected to GLC or GC-MS after reconstitution with methanol. The gastric contents are treated in a similar fashion after an initial protein precipitation step with ammonium sulphate and heat. In addition, specific examination for salicylates and certain volatiles (acetone and alcohols) must be performed as these will not be detected in the first two manipulations.

ANALYTICAL METHODS USED

Three distinct types of analytical methods are most frequently used in clinical toxicological analysis, namely (1) chromatography, (2) photometry/spectroscopy and (3) immunoassays (competitive protein binding).

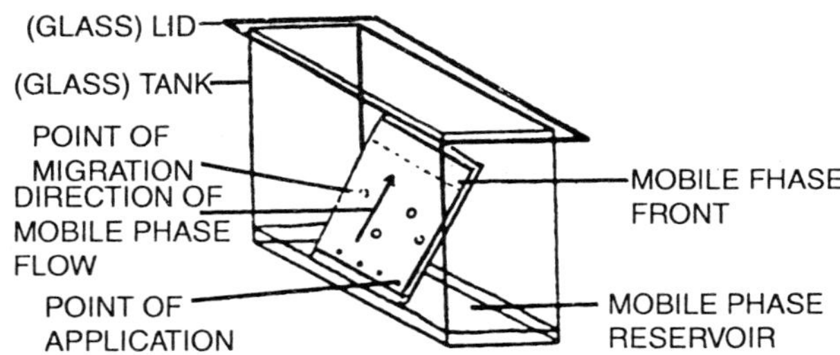

Figure 6.4 : ***Typical arrangement for thin layer chromatography.***

Chromatography

The term *chromatography* originally referred to the separation of the components of mixtures and their individual identification by specific colour development, which is, however, no longer a part of most chromatographic techniques. Common to all chromatography systems is the presence of a mobile phase (liquid or gas) carrying the mixture (patient sample), which is brought into contact with a stationary phase (liquid or solid). Separation of components is brought about by differences in the interaction between the mixture components with the stationary and sometimes the mobile phases. The types of interactions include different surface adsorption and desorption, solvent partition, molecular seiving, and ion exchange [4].

The types of chromotography may be named by (1) physical characteristics or arrangement of phases (thin-layer, column for example; (2) physicochemical mechanism of seperation (ion exchange, adsorption for example : (3) type of technique (gasliquid high, high pressure liquid for example). No nomenclature system is entirely descriptive. A thin-layer system may separate by adsorption. A high pressure liquid system may separate by ion exchange.

Thin-Layer Chromatography

In thin-layer chromatography the stationary phase may be paper or a granular adsorbent (most commonly silica) spread uniformly on an inert support (frequently a glass plate). A mixture to separated, as well as known (control) samples of its possible components, are applied as discrete spots. The spotted plate is lowered into a trough containing the mobile phase (a mixture of organic solvents) care being taken not be immerse the dried spots. The mobile phase travels up the solid phase by diffusion carrying with it mixture components. This process is performed in a closed chamber and is terminated when the mobile phase has travelled a predetermined specified distance. The separation is effected by differences in adsorption, rate of diffusion and solubility. After the solid phase is dried, specific reagents which will produce characteristic colour reactions with the agents of interest are sprayed on to the surface. Substances which naturally fluoresce or can be made to fluoresce may be visualized as well. In some instances agents of interest have natural colour which may be used as a criterion for identification.

Identification of mixture components is made by comparing the migration distance relative to that of the controls as well as by colour reaction. The R_f is a ratio of the distance a spot has moved relative to the distance the mobile phase has moved. The R_f of each unknown is compared to the R_f of each known control. The R_f is not a true constant but is system dependent. Thin-layer chromatography generally is not quantitative by itself. Separated componeuts may be eluted for chemical identification and quantitation.

Gas Chromatography

There are two modes of gas chromatography : the more frequently used gas liquid and gas solid, the use of which is limited to quantitation of gases and highly volatile compounds (anesthetics, ethanol). Only GC will be considered here.

GC is useful for the identification of organic compounds. Its high sensitivity is such that most drugs can be detected in most human samples if properly prepared. The application of GC is limited primarily by the ability to volatilize a molecular species or some derivative (see vapour pressure, below) and the thermostability of the species. For instance, carbamazepine, oxazepam, sulfonamides, propoxyphene, succinyicholine, phenobarbital, chlordiazepoxide are subject to thermolytic reactions and hence alternative quantitative methods should be selected.

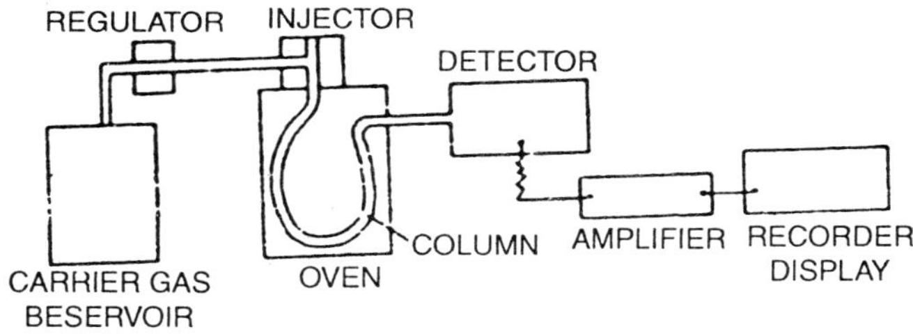

Figure 6.5 : ***Schematic of gas chromatograph basic components.***

There are eight components to a GC system: (1) carrier gas (*mobile phase*) *reservior*; (2) *regulator*; (3) *injector*; (4) *oven* ; (4) *column* (*containing stationary phase*) ; (6) *detector*; (7) *signal amplifier*; (8) *output recorder/display*. In the fundamental operation of the system the properly prepared sample is introduced into the mobile phase flow at the injector, which is at an elevated temperature to enhance vaporization of the sample. The separation takes place in the column where the sample constituents come into contact with the stationary phase, also at an elevated temperature (though lower than the injector temperature) maintained by the oven. The separted molecular species are detected as they emerge from the column by the detector. The detector's output is modified and converted into concentration by the combined effects of the amplifier and recorder/display components of the system.

In GC the mobile phase in an inert carrier gas which does not participate in reactions with sample constituents. The columns are long (meters) with narrow bores (1-2 mm) and packed with a stationary phase which is liquid at operating temperature (e.g., high molecular weight alcohols, paraffins) adsorbed onto a a support (e.g., silica). The column is contained in temperature-regulated oven of which the usual operating temperatures are 200-1,000°C. The mixture sample is introduced into the flow of gas at an injection port at which point it is volatilized by flash evaporation. Separation takes place as the mixture components remain on the stationary phase for different lengths of time according to their individual interactions with it. Compounds of similar chemical structure and relatively high boiling points are separated on the basis of chemical and physical properties. The components of the unknown mixture are partially in the stationary liquid phase and partially in the mobile phase according to their vapour pressures. The time it takes for a particular molecular species to arrive at the detector at the end of the column (the retention time) is a function of the volatility, size and polarity of the molecule, carrier gas pressure, oven temperature, and the nature of the particular stationary phase. The partition coefficient is the ratio of weight of solute per milliliter of the stationary phase to weight of solute per milliliter in the carrier gas. A compound with a high vapour pressure will have a low partition coefficient (more in carrier gas) and will elute sooner than compounds with lower vapour pressures. If there is a selective interaction between a sample component and the stationary phase, the elution order may change. Polar function groups (OH.NH, COOH) reduce the vapour pressure and so may limit the application of GC analysis. Each separated compound is detected as it is eluted. This may be accomplished by several types of detectors. The flame ionization detector (FID) is commonly used for drugs. The electron capture detector is used when greater sensitivity is needed, and is particularly useful for low level detection of halogenated compounds (e.g., chlordiazepoxide). The thermal conductivity detector is useful particularly for gases and low boiling point alcohols even though it is not as sensitive and so requires larger sample volumes.

Figure 6.6 : ***Schematic of mass spectrograph basic components.***

Mass Spectrometry:

There are two types of MS based on the way in which ionization of substances is effected : (1) *electron bombardment* (also called fragmentation) and (2) *chemical ionization*. Only fragmentation will be considered here.

MS identification is based on the degradation of test substances into ions and recording the fragmentation pattern accoming to each fragment's ratio of mass to charge (m/e ratio). In contrast to other techniques, MS is concerned with the actual molecular structure of substance rather than characteristics such as absorption or emission of light. In principle this technique produces and separates fragments, usually by electron bombardment ionization in a vacuum, and determines both the m/e ratios and the relative abundances of the fragments. The fundamental concept is that functional groups direct the fragmentation of a molecule under electron bombardment. Therefore, it is possible to predict the fragmentation pattern when the structure is known and, conversely, to estimate the structure if the fragmentation pattern is known. MS is highly sensitive as well as specific.

There are six basic functional components to the mass spectrometer: (1) inlet for introduction of sample; (2) ionizer; (3) mess separator; (4) ion detector; (5) signal amplifier; (6) output/display.

The interior of the mass spectometer is at low pressure. The inlet serves the function of allowing the introduction of the sample without the loss of the interior vacuum. A relatively pure liquid, gas or solid sample is required for introduction into the instrument. A special probe is used for introducing solid samples through a series of vaccum locks. For liquids and gases, a particularly useful inlet device is a gas chromatograph. In this respect the gas chromatograph performs the dual function of (1) sample purification and preparation (volatilization, separation of constituents) and (2) inlet for the mass spectrometer. This combination GC-MS is the most sensitive and accurate tool for the identification of organic compounds in the clinical laboratory.

Once introduced, the sample must be ionized. Electron bombardment of the vaporized sample is the technique most commonly employed. Both positive and negative ions are produced in the process, although the positive ions are produced in greater quantity. The positively charged, collimated ions are accelerated into and separated by the mass separator according to their individual m/e ratios. The three principal methods of separation are : magnetic and/or electric field deflection ; quadrupole filtering ; time of flight. Regardless of the technique, the ion beam is resolved (separated by m/e ratios) in seconds.

The detector usually is an electron multiplier. As the ions are selectively brought to impact on the detector by the mass separator, an output signal directly proportional to the number of ions of each m/e ratio is generated. The detector with the amplification system is able to produce sensitivity to the level of single ions. The ion abundance at any particular m/e ratio is registered for a very short portion of the total scan time. In a 1-second scan over a range of 500 atomic mass units, for instance, each m/e value is recorded for only 2 ms.

The display of the output is handled in several ways. The easiest way to understand the conventional mass spectrum display is to consider it as a two-dimentional plot of m/e ratios on the abscissa and relative abundance on the ordinate. The most intense peak (greatest abundance) is called the base peak and is assigned the value of 100. The intensities of all other peaks are reported as percentages of the base peak (relative abundances), This fragmentation pattern generally specific for identification. Finally, some molecules escape fragmentation and a molecular ion is recorded, allowing for the determination of molecular weight, and in some cases the empirical formula.

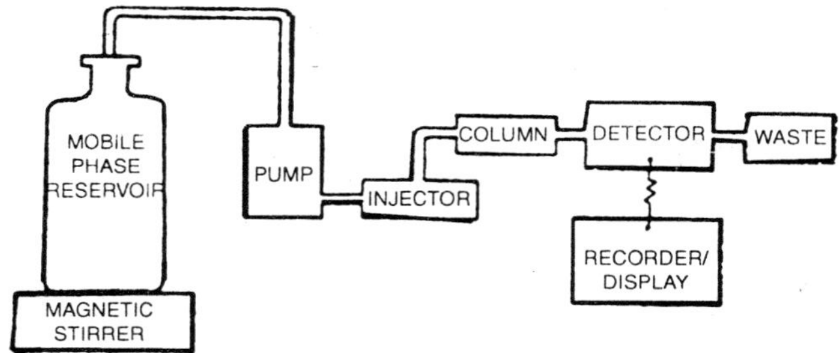

Figure 6.7 : *Schematic of high pressure liquid chromatograph basic components.*

An alternative to the conventional approach to MS is selected ion monitoring (SIM) also referred to as mass fragmentography. As has been described above, any particular m/e ratio is recorded for an extremely short period of time. For the remainder of the scan time this m/e ratio is not focused and therefore not monitored. Moreover, a significant portion of the scan time is spent between peaks during which no useful information is being obtained. In SIM only one of a very few previously selected ions are monitored during the entire process. In this mode of monitoring there is rapid switching among peak positions. The net result is an increase in the amount of time spent monitoring these present peaks (by a factor of 10 or 100), and a dramatic increase in sensitivity. The detection limit (in mass units) for SIM is much lower (greater sensitivity) than for conventional GC-MS. Of course, SIM does not yield a complete mass spectrum, for which reascn its results usually must be confirmed by other methods.

High pressure liquid chromatography (HPLC):

The basic operational pathway in an HPLC involves six elements: (1) solvent (mobile phase) reservoir ; (2) Solvent delivery system (pump); (3) sample injector; (4) column; (5) detector; (6) output data processing unit. The sample is introduced into the flow of mobile phase through an injector which is designed to operate at pressures between 1,500 and 7,000 psi. The separation takes place within the relatively short column (15-30 cm with 2-5 mm inside diameter) which is picked with the stationary phase. When a relatively polar stationary phase is used with a nonpolar mobile phase, the arrangement is referred to as normal phase; reversed phase is a relatively nonpolar stationary phase with a relatively polar mobile phase. The separation takes place due to the relative interactions of molecular species within the mixture sample with both, the mobile and the stationary phase. HPLC most often is used at ambient temperatures with permits separation of substances which might be degraded by heat or which cannot be made volatile.

Substances are identified by retention time (as in gas chromatography) under a specific set of chromatographic conditions. Quantitation is accomplished using various types of detectors, most commonly ultraviolet absorption, operating as a flow-through spectrophotometer. This detector is useful for most drugs which might be present in human materials. Fluorescence detectors are used for substances which are naturally fluorescent of those which might be made,to fluoresce after separation *e.g.* postcolumn derivatization of gentamicin with O-phthalaldehyde).

PHOTOMETRY/SPECTROSCOPY

A *photometer* measures relative radiant power or some function of it. Photometry measures light transmitting or light emitting power of a solution and thus determines the concentration of either light absorbing or light emitting substances present. This general operating principle applies to spectrophotometers, flame photometers and atomic absorption spectrophotometers which are basic to any clinical laboratory.

Spectrophotometer

A *spectrophotometer* measures the transmittance (T, the ratio of transmitted light to incident light) or absorbance (A, which is equal to log 1/T or-log T) at any one of several selectable wavelengths in ultraviolet, visible, or infrared light. If no light is absorbed (A= O) then the ratio of transmitted light to incident light is 1, and the percent of transmittance (% T) is 100% The relationship of A to % T is as follows: A = -log T = -log % T/100; since 2 is the log of 100, A = 2-log % T. A most often is measured in the scale range from 0 to 2. When adjusting the instrument to zero, one actually is setting the % T of the blank solution to 100% (*i.e.*, zero absorbance).

The straight line light path originates in a light source, is focussed on a monochromator, passes through the sample to fall on a detector, which is connected to a meter. Monochromatic light is essential; and a 'blank' solution must be used containing all constituents except the test substance to set the instrument to zero. The fundamental principle is that the absorption of monochromatic light is characteristic of a molecular species and that no other substance will absorb at specific light wavelengths in an identical pattern. Since A or %T is related to concentration in a predictable manner, it is possible to estimate concentration once either is known. The assumption is that no interfering substances (those absorbing at the same wavelength) are present.

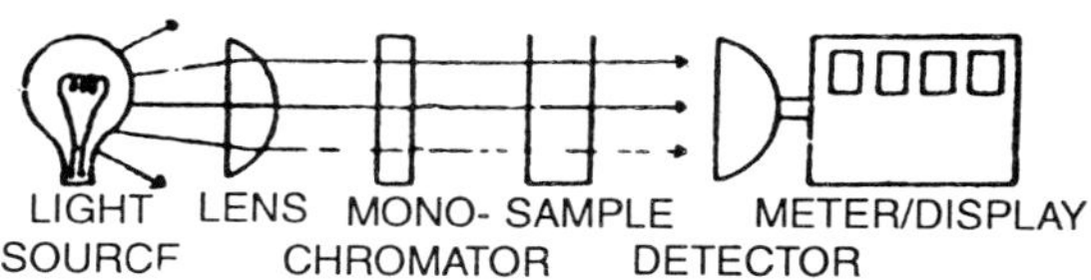

Figure 6.8 : ***Schematic of single beam spectrophotometer basic components.***

FLAME PHOTOMETER

Substances are brought to excitation in a flame and their resulting emitted light at characteristic wavelengths is quantitated in flame photometry. The basic components of a flame photometer include an atomizer, burner and fuel supply, monochromator, detector, and display. Many atoms, ions, or molecules when excited by heat emit characteristic spectra. The atom excited by thermal energy (an electron moves to an orbit of higher energy level) emits a photon as it returns to ground state (the electron returns to its lower energy level orbit). The intensity of the light emission is directly proportional to the number of excited atoms. A line spectrum is a definite, sharply delineated energy maximum of spectral emission which is essentially monochromatic. Line spectra are characteristic of molecular species and may be used for detecting and quantitating the species. The line spectrum must be isolated, so the monochromator again is the key element.

Quantitation is accomplished by the internal standard principle. A known amount of another element is added to the unknown and the two are quantitated simultaneously. For instance, it is common

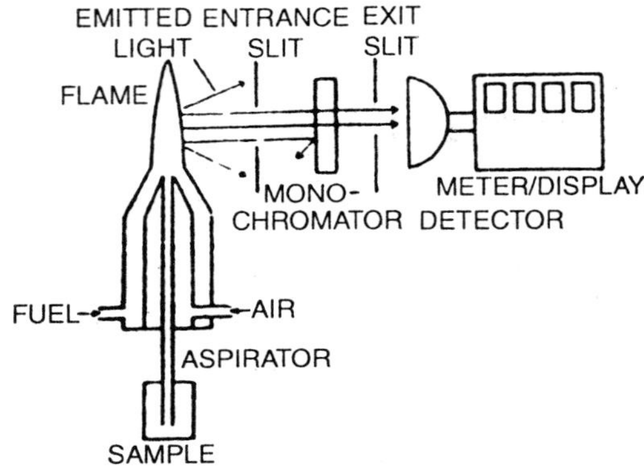

Figure 6.9 : ***Schematic of flame photometer basic components.***

to use lithium as the internal standard for sodium quantitation. The intensity of the internal standard should be the same for all standard, unknown, and blank solutions. The ratio of the intensity of the spectral emission line of the unknown is compared to that of the known internal standard and is proportional to the concentration. Obviously, the ideal internal standard is one not likely to be present in the unknown specimen.

ATOMIC ABSORPTION SPETROPHOTOMETRY

The ability of each element to absorb at the same wavelength as tbat which it emits is the basis of this technique.

Atomic absorption spectrophotometry is in this respect the inverse of emission methods (flame photometry). The atom in the ground state, the lowest possible energy form, absorbs units of energy. Ideally the atom is not excited in the flame of atomic absorption but is merely dissociated from chemical bonds and placed in an unexcited, neutral ground state. The flame may be thought of simply as the container in which the sample is being held. There are certain advantages of atomic absorption over flame photometry. The most important is greater sensitivity since there are more atoms in the ground state than there are excited atoms in the flame.

The components of the instrument are the hollow cathode light source which consists of the same element as that to be quantitated, chopper, borner, monochromator, and detector/ display. The lamp emits the characteristic spectral line by exciting the element of which it consists with thermal/ electrical energy. The emitted light energy is absorbed by the ground state molecules in the flame. The energy of the absorbed radiation is equal to that

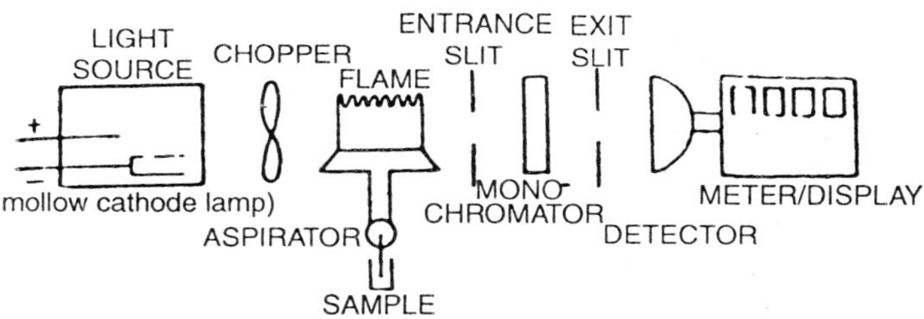

Figure 6.10 : ***Schematic of atomic absorption spectrophotometer basic components.***

which would be emitted if the molecules were excited, and is related to the concentration of the molecules. A potential problem is interference due to the fact that some few atoms are in fact excited in the flame and that the element emits at the same wavelength that it absorbs. This problem is overcome by introducing modulation of the light source signal either mechanically or electrically. The chopper (inserted between light source and the flame) is a mechanical means of converting what is a direct current (DC) signal of the lamp to an alternating current (AC) signal.

Electrical pulsing of the light source is another means of modulating the signal. The detector is then synchronized to the pulsation and rejects the unpulsed signal originating in the flame. By this manipulation, only pulsed light emerging from the flame is used fot quantitation.

Immunoassays

All protein binding assays rely on the interaction between the analyte and a specific analytical reagent to which it binds. The specific analytical reagent is a binding substance which first is specific in relative terms only, and second is one of a group of substances including receptors, transport proteins, enzymes, and antibodies. These and other binding substances have characteristics which have been exploited for a variety of purposes. From the assay point of view, antibodies as reagents differ from the other binding substances only by virtue of their mode of manufacture. The term immunoassay is used in a broad generic sense when an antibody is involved. Likewise, there are terms such as receptor assay, protein binding assay, etc.

The immunoassays are named according to the mode of labelling used for detection and quantitation purposes. Either the substance which is to be measured (the analyte) or the binding substance may be labelled. Competitive protein binding (sometimes called saturation analysis) is the generic term applied to assays in which the analyte is labelled. In competitive assays, the amount of binding substance (antibody, etc.) is kept limited well below that which is needed to bind all ligand present. The antibodies of immunoassays confer the sensitivity and specificity of the method while the labelled analyte is used to monitor the binding reaction. Sensitivity is the result of the binding affinity of the antibody for the antigen. Specificity is the result of the uniqueness of the fit of the antibody and its binding site on the antigen.

Immunoassays may be considered as being in two categories : heterogeneous; and homogeneous. They differ in that either the bound or the unbound phase of the assay system must be separated. Another important difference is that in heterogeneous assays the label is passive in the sense that its sole function is to identify the location of the molecule to which it is attached. The label (or its signal) is not modified in heterogeneous assays. Labels which have been used include radioactive isotopes, fluorescers, chemiluminescent substances, organometal groups, erythrocytes, latex beads, and enzymes. The label in homogeneous assays by contrast are active in the sense that there is modification of the output signal depending on whether the molecule to which it is attached is bound or unbound. Labels which have been used include fluorescers, stable radicals, enzymes, haptens, enzyme substrates, coenzymes, enzyme inhibitors, bacteriophages, radioisotopes, nonenzymatic catalysts, and heavy metals.

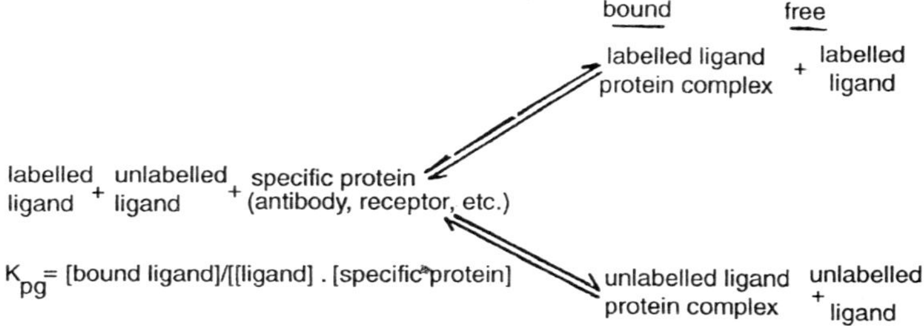

Figure 6.11 : *Principle of immunoassays.*

The obvious advantage of homogeneous systems is their relative operational simplicity. Precluding a physical separation step reduces a source of error and decreases the test performance time. As a practical matter, however, there are limitations to certain applications in the clinical laboratory setting.

RADIOIMMUNOASSAY (RIA)

RIA is a well-established clinical laboratory technique which may be applied to the quantitation of any substance to which an appropriate antibody may be produced. It offers sensitivity (down to 10^{-19} mol) beyond that achieved by chromatographic techniques. RIA as usually performed in the clinical laboratory is an heterogeneous competitive protein binding immunoassay (the analyte is labelled).

The components needed to perform RIA include: (1) the substance to be measured as supplied in the patient sample (the analyte, referred to as the ligand); (2) radiolabelled ligand supplied by the test system; (3) antibody specific for the ligand; (4) a method for separating the ligand bound to specific antibody from the unbound (free) ligand; (5) an instrument to quantitate radioactive counts per unit of time. The substance to be measured (the unlabelled ligand) in the patient sample competes with the radiolabelled ligand for the finite number of binding sites on the antibody component of the system. When an antigen is incubated with its corresponding antibody, immune complexes are formed.

This results in bound (to antibody) ligand (labelled and unlabelled), and unbound (free) ligand (labelled and unlabelled). In a reaction of this type, either the rate or the extent of the reaction may be monitored. The rate is not as reproducible as is the extent in RIA. The extent may be expressed as a constant K_{eg} which equals the concentration of the complexes (antibody bound to ligand) divided by the product of free antigen (free ligand) and free antibody. The value of K_{eg} is an expression of antigen/antibody affinity. Sensitivity is related to K_{eg}. The higher K_{eg} is,the smaller is the concentration of substance which may be quantitated. The reciprocal of K_{eg} approximates the ideal condition sensitivity of the assay. Under ideal conditions the percentage of unlabelled ligand bound to antibody is directly proportional to the total amount of unlabelled ligand present originally in the patient sample. As more unlabelled ligand is added, within limits, more of it will become bound to the fixed number of binding sites. Likewise, it should be apparent that there is an inverse relationship between the bound labelled ligand component and the bound unlabelled ligand component. After separation of the bound and free components, the radioactivity (counts per unit of time) of either fraction may be determined. Along with the patient samples, there are standards (usually $_5$ or 6) of the unlabelled ligand run. A curve relating the radioactive counts per unit time of the standards to their known concentrations is constructed. Then by locating the counts per unit time of the unknowns (patient samples), the corresponding concentration is derived. In fact there are many different possible mathematical manipulations of the data. Most commonly either the percent bound (% B), percent free (%F) or the ratio of the counts of each component (B/F) is plotted as counts per minute versus concentration. The specificity of the technique is defined by the antibody. For this reason RIA (as well as other immunoassays) may be used to confirm results of another technique. For instance, chromatographic techniques depend on relative retention times, assuming no interfering substance with a similar retention time is present. In some situations it is desirable to confirm the identity of a chromatogram peak. Likewise, there may be some immunologic cross-reactivity among similar molecular structures, requiring confirmation of immunoassay results by alternative techniques.

ENZYME IMMUNOASSAY (EIA)

In ETA the label is a nonradioactive enzyme. The enzymes used include horseradish peroxidase, lysozyme, glucose oxidase of *Aspergillus niger*, alkaline phosphatase of *Escherichia coli*, or veal intestine, and galactosidase of *Escherichia coli* glucose-6-phosphate (G-6-P) dehydrogenase of *Leuconostoc mesenteroides*, and malate dehydrogenase.

EIA may be heterogeneous or homogeneous. There are many acronyms for heterogeneous EIA. The term enzymelinked immunosorbent assay (ELISA) is best recognized as referring to heterogeneous EIA. An obvious requirement is that the labelled analyte be immobilized. Plastic (polystyrene) and cellulose have been used as well as other materials. The solid phase material may be the test tube wall or beads.

The homogeneous EIA has been exploited for use in the clinical laboratory to a greater extent than has the heterogeneous assays. Enzyme multiplied immunotechnique (EMIT, Syva Corp., Palo Alto, Calif.) is a homogeneous assay, and is the one which presently is the most common of all EIA (heterogeneous or homogeneous) in practical use in the clinical laboratory.

EMIT assays are based on the same competitive binding principles as RIA. The difference is that an enzyme label is used rather than a radiolabel. In the quantitation of drug, for instance, the ligand analyte is labelled by conjugation with bacterial G-6-P dehydrogenase (1.1.1.49) in such a way that the enzyme's activity is not significantly altered. The enzyme bound to drug converts substrate G-6-P to gluconolactone-6-P (G-6-PD) and coenzyme NAD+ to NADH which is monitored at 340 nm. First the patient sample containing the unlabelled ligand is incubated with a reagent containing antibody specific for the ligand, substrate for the enzyme, and the NAD+ coenzyme. The labelled ligand is added subsequently as a second reagent which is appropriately buffered.

The labelled ligand combines with the remaining antibody binding sites. The binding of antibody to the drug reduces enzyme activity and decreases the production of NADH. When antibody binds to the labelled ligand, the enzyme activity is inhibited due to blocking of its active site. Therefore, the more digoxin, for example, is in the patient sample (analyte, unlabelled ligand), the more antibody will be bound to it, the less antibody will be bound to the labelled digoxin, and the more NADH will be produced. Therefore, the residual enzyme activity is directly proportional to the amount of unlabelled ligand (digoxin) originally present in the patient

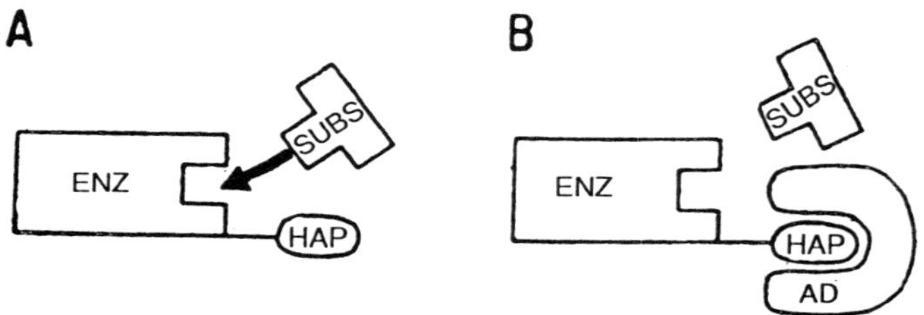

Figure 6.12 : ***Principle of homogeneous enzyme immunoasay (EMIT).***
A. The substrate (SUBS) has access to the enzyme (ENZ) when no antibody (Ab) is bound to the analyte hapten (HAP).
B. When the analyte hapten is bound to an anti-body no enzyme substrate interaction takes place.

sample. The more drug in the patient sample, the more NADH will be produced. The bacterial coenzyme will not interact with human G6-PD, so that there is no interference from that source. The principal operational difference between this homogeneous assay and RIA is the lack of a separation step. The advent of EMIT assays has facilitated the confirmation of chromatographic techniques as well as the introduction of therapeutic drug monitoring and drug screening into general clinical laboratories as the technique is compatible with the type of equipment usually already available.

FLUORESCENT IMMUNOASSAY (FIA)

Fluorescent molecules are characterized by the ability to absorb light at specific wavelengths (usually ultraviolet and visible wavelengths) and emit lower energy, longer wavelength light in response. The time delay is approximately 10^{-9} 10^{-4}s. There are many factors which influence the theoretical as well as the practical limits of sensitivity for FIA. The competitive protein binding principles are the same as for RIA and FIA. Likewise, heterogeneous as well as homogeneous assay systems have been used.

Heterogeneous FIA has the advantage of reducing interference from fluorescence intrinsic to the sample matrix and/or the sample container. This has the practical result of increasing sensitivity, at least theoretically. There are four configurations to heterogeneous FIA:

(1) Indirect, usually a serologic application, wherein antibody in the patient sample (the analyte in this case) binds during incubation to immobilized antigen. Subsequently fluorescentlabelled antibody which will bind to the residual antigen sites is added. The fluorescence on the solid phase is indirectly (from where this configuration derives its name) related to the analyte concentration in the original sample.

(2) Competitive, wherein labelled and unlabelled analyte compete with a fixed, limited number of binding sites on antibody immobilized on the solid phase. The fluorescence on the solid phase is then quantitated.

(3) Sandwich, wherein antibody to the analyte is immobilized on the solid phase. After incubation with the sample containing the analyte, a second antibody which is fluorescent labelled is added. The fluorescence on the solid phase is directly related to the analyte concentration.

(4) Fluoroimrnunometric, wherein the analyte competes for a fixed, limited number of antibody binding sites in a solution. Note that in this configuration the antibody is labelled, not the analyte. The residual unbound labelled antibody later binds to solid phase bound antigen in a subsequent incubation. The fluorescence on the solid phase is indirectly proportional to the analyte concentration in the sample.

The two solid phase, heterogeneous FIA systems in use in the clinical laboratory at this writing are the FIAX system which uses a flat polymer on a plastic paddle called a StiQ (IDT, Santa Clara, Calif.) and the Fluoromatic system which uses transparent suspendable polyacrylamide microbeads (BioRad, Richmond, Calif).

The homogeneous FIA systems are : (1) substrate-labelled homogeneous fluorescent immunoassay; (2) fluorescence polarization; (3) fluorescence excitation transfer immunoassay; (4) fluorescence protection immunoassay (FPIA)-none of which are in wide use in the clinical laboratory. Substratelabelled homogeneous fluorescence, and fluorescence polarization immunoassays as techniques for drug quantitation are just being introduced at the time of this writing.

The substrate-labelled fluorescence immunoassay is a twostep process based on specific properties of a conjugate of analyte to fluorogenic dye. The nonfluorescent-labelled analyte serves as a substrate for an enzymic reaction which yields a fluorescent product. The nonfluorescent-labelled analyte must also be capable of binding to an analyte-specific antibody. The antibody bound complex must have the characteristic of not acting as a substrate for the enzyme reaction. In the first step, the competitive binding reaction is set up by having sample containing an unknown amount of analyte incubated with a mixture containing nonglvorescent-labelled analyte and a limited amount of analyte-specific antibody. The amount of nonfluorescent-labelled analyte which has not bound to antibody is determined by the amount of analyte originally present in the sample with which it competes and may be quantitated in the second step. The enzyme is added as the second step after the competitive binding reaction is at equilibrium and results in the production of fluorescent product from the nonfluorescent-labelled unbound analyte. The fluorescence in this respect is directly related to the concentration of analyte originally present in the sample.

An example of a substrate-labelled fluorescent immunoassay is the AMES TDA (Ames Division, Miles Laboratories Inc., Elkhart, Ind.). In this technique the enzyme substrate is a derivative of the fluorescent dye umbelliferone which is covalently joined to galactose via an acetyl linkage. The analyte is then coupled to the carboxyl group of this complex. The analyte linked to the umbel liferyl- B-Dgalactoside is nonfluorescent and acts as the substrate for galactosidase. By hydrolysis, the substrate-drug complex becomes fluorescent.

Fluorescent solutions emit partially polarized wavelengths when viewed at right angles to an excitation beam of vertically polarized or natural light. This fluorescence polarization is due to the fixed relationship between molecular orientation, light absorption and fluorescence emission. The degree of polarization is inversely related to the speed of rotation (tumbling) of the molecule in solution.

Fluorescence polarization takes advantage of the fact that small molecules rotate rapidly and have low polarizations and large molecules rotate slowly and have high polarizations. Therefore, when a low molecular weight fluorescent ligand combines with a large receptor molecule (such as antibody) the fluorescence rises in response to the decrease in velocity of the rotation (rotatory Brownian Motion).

TOXICOLOGICAL TESTING PROCEDURES

The toxicity is generally put into following types:

1. *Acute toxicity* : Toxicity caused by a high dose at short term exposures (suddenly).

2. *Chronic toxicity* : It refers to the toxic effects caused by high or low dose at long term exposure. 3. *Lethal* : Causing death.

4. *Sublethal* Below the level which causes direct death.

5. *Cumulative* : Increase in the toxicity level or reaching to toxic level by gradual addition of toxicant's quantity. Following types *of* tests are in practice.

Short Term Acute Toxicity Tests

In this test, the organisms are exposed to various concentrations to toxicant for a short duration of 24, 48 or 96 hrs. and the mortality is noted LC_{50} (lethal concentration 50) is calculated for toxicant or pollutant, which means the concentration at which 50% of the organisms are killed in a specified time. If mortality is not a criterion EC 50 (affective concentration 50) is calculated which means the concentration which brings desired effect in 50% population.

Life Cycle Toxicity Tests

This test involves exposing several groups of individuals of one species to different concentration of a toxic agent through part of a life cycle in order to study the effect of the toxic agent on survival, growth and reproduction. Partial life cycles tests are conducted with fish species that require more than a year to reach sexual maturity. In this case the test can be completed in less than 15 months, still exposing all major life stages to the toxic agent begins with immature juveniles at least 2 months prior to the active gonad development continues through maturation and reproduction and ends not less than 30 days (90 days for salmonids) after watching of the next generation.

Life Cycle Toxicity Test

This test begins with employes of newly hatched larvae less than 48 hrs. old, continues through a maturation and reproduction stage and with fish ends not less than 30 days (90 days for salmonids) after the hatching *of* the next generation.

Both these tests are used in the calculation of maximum acceptable toxicant concentration.

Maximum acceptable toxicant concentration

It is the highest concentration of a toxicant that has no adverse effect on survival, growth and reproduction of a species based on the result of a partial life cycle toxicity test. However, a life cycle or partial life cycle test can not produce an exact value for the MATC, it can only produce limits within when the MATC must fall. It can also be determined by the use of application factor,

$$\text{Application Factor} = \frac{\text{MATC}}{\text{LC}_{50}\,96\,\text{Hrs.}}$$

This application factor can be determined for one toxicant and can be extrapolated to various fishes. It varies from 2 to 5 for different fish species for one toxicant. It is however, assumed that it is constant for one toxicant.

Bioassay Test

A bioassy test is conducted to find out the toxicity of a substance. Although multi-species toxicity testing is increasingly becoming popular. Bioassay test is usually conducted with one species. Algal, higher plants, Zooplankton and fish bioassay tests are described here. The selection of suitable organisms for bioassay tests depends on following factors (Jenkins 1981) (APHA 1985).

1. The organism should be sensitive to the toxicant.
2. It must be capable of accumulating the substances (for long term tests).
3. It must be common in occurrence, be geographically widespread.
4. It must be readily available in good numbers throughout the year.
5. It should have economic, recreational or ecological importance both locally and nationally.
6. It should be easily cultured in the laboratory.

Fish Bioassays

'Bioassay' is a test in which organisms are used to detect the presence or the effects of any other physical factor, chemical factor or any other type of ecological disturbances. The bioassays are extremely common in water pollution studies and are also used by enforcement authorities for regulating the discharge limits or the safe concentrations of various waste and chemical substances. However, the bioassay can also be taken as a tool for identifying the effect of or tolerance to a chemical by organisms. The bioassay thus could find a very important role in studying synergistic effect of chemical, assessment of efficiency of a waste treatment method etc.

The bioassay can be conducted by using any type of organisms, however, the fish bioassays are very common and useful.

The aim of bioassay is to find out either lethal concentration or effective concentration causing mortality or other effects. Ultimately they are to be utilised for determination of safe concentration of a chemical or maximum acceptable toxicant concentration.

In the bioassay test the organisms are exposed to different concentrations to a toxicant for a definite period and mortality, behaviourial change or other signals of distress are noted periodically.

Types of bioassays

The bioassay can be divided into three types based on the change in toxicant solution.

1. *Static Bioassay*

In this bioassay test, the arrangements are not provided for renewing or changing the test solution during the test period, i.e., the organisms are exposed to the same toxicant solution for the whole experimental period.

2. *Renewal Bioassay*

In this type of test the test solution may be changed many times during the experimental period.

3. *Flow-through Bioassay*

The Flow-through Bioassay are those in which an arrangement for continuous flow of test solution is made, *i.e.* the organisms are exposed to flowing water similar to a natural situation.

A Static Bioassay procedure is described here. It should be noted that this type of bioassay test can not be made applicable to all kind of toxicants studies. Its application can be made:

1. To study the efficiency of various methods of wastewater treatment.
2. To monitor the impact of any waste discharge on any aquatic ecosystem.
3. To screen various chemicals like pesticides, synthetic organic chemicals etc. for their deleterious environmental effects at tentative level, to study the relative toxicity of various chemicals.

Calculation of LC_{50} EC_{50}, or TLm

The results are recorded on a semilong graph paper with the concentration being recorded in long scale and the mortality or any other effect in arithmatic scale.

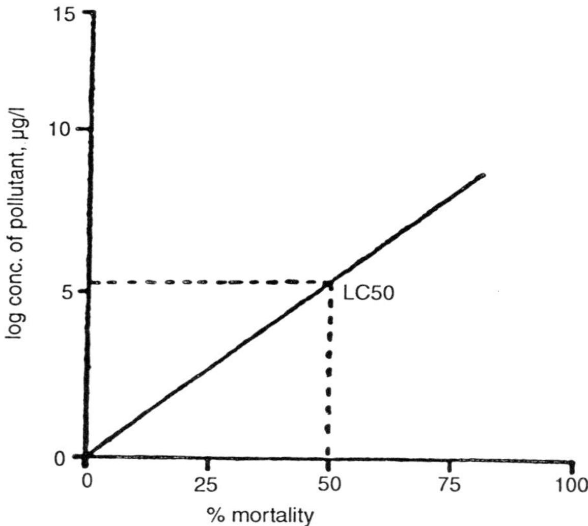

Figure 6.13 : ***A Simple toxicity curve.***

A line is drawn from 50% lethality level to the graph line and is connected to the concentration. This is LC_{56} EC_{50} or TLM for a particular period like LC_{50} 24 hrs, or LC_{50} 96 hr.

While reporting the bioassay results a mention must be made of the species of the fish used and their length, weight, pH, DO, as well as temperature etc. at which the test has been carried out.

7

Soil Analysis

The soil consists of five major components: mineral matter, organic matter, air water and micro organism. Mineral matter forms the bulk of soil (more than 90% by weight). Almost all the elements found on the earth are present in the soil, however, most of them occur in trace quantities.

Soil consists of different particulates, like stones, large pebbles, coarse sand, fine sand, silt, clay, humus and some organic matter. As per the percentage of silt and clay, the soil can be classified into various classes as follows

Soil class	*Silt and clay percent*
Sand	Less than 7
Loamy sand	7-15
Sandy loam	16-25
Light loam	26-40
Heavy loam	More than 40

COLLECTION

The surface sediment samples from the water bodies are collected with the help of dredges. The most common in use is the Ekman dredge. To sample the deeper layers in aquatic systems, special kind of boring machines are employed. Soon after the collection of soil/sediment samples, they are transported to the laboratory and are spread for air drying. For some parameters, it is essential to use the fresh samples, while for others, the soil can be stored variously after proper dryirg.

The drying temperature for soil is critical as the nature of many constituents is highly sensitive to temperature. High temperature drying generally affects the exchangeable characteristics of clay and organic colloids.

After proper drying of the soil, large stones and other similar objects are removed and the soil is ground to break up aggregates and crumbs, taking care not to break actual soil particles. The soil is then worked through a 2 mm screen which allows all the naturitionally important fractions to pass through. The screens made up of nylon or stainless steel should be preferred over the brass as the latter may add zinc and copper to the soil.

PARTICLE SIZE ANALYSIS

Relative proportion of the soil particles of various sizes is an important physical parameter which determines texture of soil. Larger particles help in providing the physical support to the plants, while smaller size particles determine the capacity of soil to hold the water and availability of nutrients.

According to the International system, the range of the particle sizes in the soil is under

Coarse sand	0.2-2.2 mm
Fine sand	0.02-0.2 mm
Silt	0.002-0.02 mm
Clay	<0.002 mm

The percentages of the above particles can be measured by several methods such as (1) Pipette method (2) Beaker method, (3) Hydrometer method and (4) Coulter counter method.

The most simple and quicker method of the all is hydrometer method. The method is suitable only when the organic matter, salts and calcium carbonate are present is low amount. In case of the presence of these substances in appreciable quantities, the pretreatment of the soil is required. The pretreatment to the soil is given as follows :

In case of more than 12% organic matter, it should be removed by hydrogen peroxide treatment. For this, take 50g of soil, cover with water and add 20 ml H_2O_2. Heat until the reaction ceases. If necessary, repeat the process until all the organic matter is destroyed. To remove excess $CaCO_3$, add slowly 1 + 1 HCl to 50g of soil, warm and stir until reaction subsides. Continue the addition of HCl until all the $CaCO_3$ is destroyed. Filter the soil and wash to remove all the excess acid. For saline soils, wash them thoroughly to leach the salts out. Dry the soil and reweigh before proceeding.

Standard hydrometer with Bouyoucos scale in g/L, electrically driven Stirrer and graduated. Cylinder of one litre capacity and the required apparatus.

Reagents:

(A) Calgon, 5%

Dissolve 50 g of Calgon in distilled water, adjust the pH at 9 with (Na_2C0_3) and make up the final volume to 1 litre.

Procedure (for Silt and clay)

Weigh 50 g of air dried 2 mm soil two times and transfer one portion to a 1 litre capacity beaker. Keep the second portion of 50 g of soil in the oven at 105°C for overnight to find out the moisture content of it Add 25 ml of 5% calgon and 400 ml of water to the beaker soil. Allow the sample to soak for at least 10 minutes.

Mix the suspension for thorough dispersion by stirring for 15 minutes with the help of an electrically driven high speed mixer. Transfer the contents to a 1 litre graduated cylinder removing all the adhered soil with the help of a stream of water, and bring the level to the mark of 1 litre with water.

Stir for 1 minute with a paddle or a glass rod. Record **the** timing for readings with the soil hydrometer. Add a drop of amyl alcohol in case the surface is covered with after stirring. Take the reading according to the following scheme, introducing the hydrometer just 20 sec. before reading.

(1) 4 min. 48 sec = International silt and clay (<20,u)

(2) 5 hours = International clay (<2µ)

Take the temperature of suspension after each reading.

Calculation

As normally the *Bouyoucose hydrometer* is calibrated for a temperature of 19.5°C, a temperature correction in the readings is required. For temperature correction, 0.3 units are either added or subtracted for every degree above or below 19.5°C respectively, in the hydrometer readir.g.

$$\text{Clay \%} = \frac{A \times 100}{S - M} - 1$$

$$\text{Silt + Clay\%} = \frac{B \times 100}{S - M} - 1$$

(1 is substracted as calgon correction)

Silt% = (silt + clay)% – clay%.

Total Sand % (approx) =100 – (Silt+clay)%

where,

A = Hydrometer reading (g/L) after 5 hours.

B = Hydrometer reading (g/L) after 4 min. 48 sec.

S = Weight of soil taken (50 g)

M = Moisture content in 50 g of soil (in g)

Procedure (for fire and coarse sand)

After second reading of hydrometer (as for clay above) decant all the supernatent. Add more water upto the mark of 1 litre, and again discard the supernatent. Repeat the procedure until the supernatent liquid is clear.

Transfer the remaining portion of the soil into a BS 72 mesh sieve supported preferably over a funnel above a beaker. Wash the soil over the seive with gentle rubbing. Dry the sieve in an oven at 105°C for 3 hours, transfer the soil in a preweighted basin, cool in a desiccator and take the weight. Similarly, transfer the fine sand from the beaker to a basin, evaporate to dryness, heat in an oven at 105°C for 3 hours, cool in a desiccator and weigh.

Calculation

$$\text{Coarse sand \%} = a \times \frac{100}{S - M}$$

$$\text{Fine sand \%} = b \times \frac{100}{S - M}$$

where,

a = weight of coarse sand in g (retained on the sieve)

b = weight of fine sand in g (passed through the sieve)

S = weight of soil taken (50 g)

M = Moisture content of 50 g soil (g)

WATER HOLDING CAPACITY

Inter holding capacity of soil usually refers to amount of maximum water which can be held in the saturated soils. It is generally measured as the amount of water taken up by unit weight of dry soil when immersed in water under standardized conditions.

Requirements

Circular. Brass Boxes:

These are normally of the dimensions of 5.6 cm internal diameter and 1.6 cm height. The bottom is perforated with several holes (0.75mm in diameter) spaced at 4 mm. Sometimes, however, instead of circular boxes, square boxes can also be used with more or less same volume.

Procedure:

Place a filter paper in the brass box (Whatman No. I or 44) of an appropriate dimension so as to cover the whole perforated bottom of the brass box. Take the weight of brass box plus filter paper (W_1). Now transfer the 0.5 mm dry soil in small portions, tapping the box gently after each addition, until the box is nearly full. Then add sufficient soil to fill the box and strike off the extra soil level by moving one spatula over it.

Place the packed brass box in a petri plate and add water upto a depth of about 1 cm. After some water has been absorbed by the soil, restore the depth of water in petri plate by addition of some more water and keep for overnight.

Next day after 12-16 hours, remove the box, wipe it out dry from outside and take the weight (W_2). Now dry it in an an oven for about 24 hours at 105°C, cool in a desiccator and again weigh (W_3).

Also find out separately the amount of the water absorbed by the filter paper. For this, weigh a few filter papers (say for example 5 pieces), and after saturating them with water, again take their weight. Calculate the amount of water absorbed by a single filter paper (W_4).

Calculation:

W_1 = Weight of brass box+filter paper

W_2 = Weight of brass box+saturated soil

W_3 = Weight of brass box+ oven-dry soil

W_4 = Amount of water retained by the filter paper

$$\% \text{ Water holding capacity} = \frac{W_2 - W_3 - W_4}{W_2 - W_1} \times 100$$

TEMPERATURE

The temperature of soil is more variable at a surface layer, but ceases to show much variation with the greater depths.

Requirement

Soil Thermometer

This is actually a mercury thermometer, the bulb of which is kept inside of a metal cone to facilitate the penetration into the soil. The upper side of a metal cone is fixed with a wooden shaft which supports the stem of the thermometer to see the reading.

Procedure

Insert the cone of the soil up to desired depth and note the reading directly on the stem which is open in the air.

pH

pH of soil is the measure of 'hydrogen ion activity' and depends largely on the relative amounts of the adsorbed hydrogen and metallic ions. It is a good measure of the intensity of acidity and alkalinity of a soil-water suspension, and provides a good identification of the soil chemical nature.

pH of soil suspension highly depends on the soil water ratios. Determination of pH at moisture saturation level and 1 : 5 soil suspension is most common.

Procedure

To determine the pH at the moisture saturation level, take about 50 g of soil in a beaker. Add small portions of distilled water, without stirring the soil, until a glistering layer appears at the surface. Now make a uniform paste of soil by stirring with the help of a glass rod.

For determination of pH of soil in 1 : 5 soil suspension, take 20 g of soil and add 100 ml of distilled water. Stir for about an hour at regular intervals.

Determine the pH electrometrically by using pH meter in unifiltered soil solutions.

Results are expressed directly in pH units associated with the specific dilution of soil suspension, e.g., pH in 1 : 5 soil suspension.

CONDUCTIVITY

Conductivity, as the measure of current carrying capacity, gives a clear idea of the soluble salts present in the soil. It is conventionally determined in 1 : 5 soil suspension. Like water the conductivity of the soil suspension is also measured by a conductivity meter.

Apparatus

Conductivity meter, thermometer, glassware to prepare soil suspension.

Procedure

To prepate 1 : 5 soil solution, take 20 g soil in 100 ml of aerated distilled water and shake mechanically for about an hour. Measure the conductivity of the suspension within one hour of preparation of soil suspension by directly dipping the conductivity cell into the suspension (see water analysis Chapter 5. Take the temperature of soil suspension to convert the results at 25°C employing the Table given for water analysis.

Calculation Expression of Results

Conductivity at 25 °C = observed conductivity cell constant
temperature factor

Conductivity is generally expressed in mmho or μ mho (a recent unit equivalent to mho is Seimens, S). The results should always be accompanied by the specify dilution of the soil suspension, e.g., conductivity mmho at 25° in I : 5 soil suspension.

CALCIUM AND MAGNESIUM

Cations present in the exchange complex of the soils can be removed by leaching the soil with ammonium acetate solution. Different exchangeable cations are then estimated separately in this ammonium acetate leachate.

For determination of the exchangeable cations, the soil is first washed with ethyl alcohol to remove the soluble fraction. However, for total cations (exchangeable + soluble), there is no necessity of prior alcohol washing. For soluble fraction only, a 1 : 5 soil solution is used to estimate the cations.

Reagents

A. Ethyl alcohol, 40%

Mix 600 ml of distilled water with 400 ml of absolute or 95% alcohol.

B. Absolute alcohol.

C. Ammonium acetate solution.

Dilute 57 ml of glacial acetic acid to 800 ml with distilled water and the neutralize to pH 7.0 with concentrated NH_4OH. Make up the final volume to 1 liter.

D. Aqua-regia

Mix 3 parts of Conc. HCl with 1 part of Conc. HNO_3.

E. All other reagents used in determination of calcium and magnesium in water.

Procedure

Take 50 g of air dried soil in a 500 ml beaker and add about 100 ml of 40% alcohol. Shake well and keep for about 15 minutes. Filter the suspension through *Whatman No. 50* filter paper using *Buchner funnel* and *vaccuum pump*. Wash the soil 4-5 times with 50 ml portions of 40% alcohol and perform the final washing with absolute alcohol to dry the soil. Remove the filter paper and scrap the soil in a 250 ml beaker and wash finally the Buchner funnel and filter paper with 100 ml ammonium acetate solution to remove and adhered portion of the soil. Stir the soil suspension and keep for overnight. Now filter the supernatent and later the soil with additional ammonium acetate through Whatman No. 42 filter paper, using Buchner funnel. Leach the soil 4-5 times more with portions of ammonium acetate and make up the final volume of the filtrate to 500 ml with distilled water in a volumetric flask.

Ammonium acetate and dispersed organic matter when present in appreciable amounts, interfere with the titration with EDTA ; they must almost entirely be removed prior to titration with EDTA.. Evaporate a portion of the ammonium acetate extract to dryness the dissolve the residue in a very small portion of aqua-regia. Evaporate again to dryness and dissolve the residue this time in distilled water make up the original volume of the extract evaporated.

Find out the concentration of calcium and magnesium following the method described for the analysis of water.

In case total cations (exchangeable + soluble) or when soluble fraction is very less, do not apply the alcohol washing and add directly the 100 ml of ammonium acetate to 50 g of Soil and proceed further.

Calculation

$$\%\text{Calcium} = \frac{A \times 400.8 \times V}{v \times 1000 \times S}$$

$$\text{Calcium, meq} / 100\text{g} = \frac{A \times 400.8 \times V}{v \times 20.04 \times 10 \times S}$$

$$\%\text{Magnesium} = \frac{(B - A) \times 400.8 \times V}{v \times 10000 \times S \times 1.645}$$

$$\text{Magnesium, meq / 100g} = \frac{(B - A) \times 400.8 \times V}{v \times 10 \times S \times 1.645 \times 12.16}$$

where,

A = Volume of EDTA (ml) used for calcium determination

B = Volume of EDTA (ml) used for calcium +magnesium determination.

V = Total volume of soil extract prepared (in this case 500 ml).

S = Weight of soil taken (in this case 50 g). v=Volume of soil extract titrated (ml).

Values can also be converted in mg/100 g (mg percent) after multiplying the % values by 1000.

SODIUM AND POTASSIUM

Exchangeable sodium and potassium can also be determined in ammonium acetate leachate by flamephotometric method.

Procedure

Use the remaining ammonium acetate leachate of soil after

calcium and magnesium determination for sodium and potassium. Find out the Na and K by flamephotometry as described in case of water analysis.

Calculation

$$\%\text{Na} = \frac{\text{mg Na / L of soil extract} \times V}{10000 \times 4}$$

$$\text{Na meq / 100g} = \frac{\text{mg Na / L of soil extract} \times V}{10 \times S \times 23}$$

$$\%\text{K} = \frac{\text{mg K / L of soil extract} \times V}{10000 \times S}$$

$$\text{K, meq / 100g} = \frac{\text{mg K / L of soil extract} \times V}{10 \times S \times 39}$$

where,

V = Total volume of soil extract prepared,

S = Weight of soil taken.

To convert the values in mg/100 g (mg percent) multiply the % values by 1000.

PHOSPHORUS

Phosphorus in soils is generally determined as available phosphorus, which can be extracted from soil with (002 NH_2SO_4 (1 Soil: 200 H_2SO_4). It is always advisable to measure P in fresh soils, however, if the soils are to be preserved air dry them instead of oven drying to minimise the error.

Reagents:

(A) Sulphuric acid, 0.002 N

Prepare a stock solution of 0.1 N H_2SO_4 by diluting conc. H_2SO_4 360 times (2.78- 1000 ml) and standardize with an alkali. A desired volume of this stock solution is diluted 50 times to obtain 0.002 N $\mathbf{H_2SO_4}$. Add 3 g of ammonium sulphate, $(NH_4)_2$ SO_4 or potassium sulphate, K_2SO4 per litre to produce a final pH of 3.0.

(B) All other reagents used in determination of inorganic phosphorus in water.

Procedure:

Take a 2 mm fresh soil and determine its moisture content. The moisture content is the difference in weight between fresh and after its drying in oven at 105°C. Now take fresh soil equivalent to 1.0 g oven dry weight in a 500 ml conical flask and add 200 ml 0.002 N H_2SO_4. In case of air-dried soil, directly 1.0 of soil is taken. Shake the suspension atleast for 1/2 hour. Filter through Whatman No. 50 filter paper to get a clear solution. Find out the concentration of phosphorus following the method given for determination of inorganic P in water.

Calculation:

$$\%\text{ available P} = \frac{\text{mg P / L in soil solution}}{50}$$

$$\text{P mg / 100g} = \frac{\text{mg P / L in soil extract} \times 1000}{50}$$

NITROGEN

Nitrogen in the soil is present mainly in the organic form together with small quantities of ammonium and nitrate forms. The most common method for finding out nitrogen in soil is Kjeldahl method, which measures only organic and ammonium forms excluding nitrate.

Apparatus and Reagents

A Distillation assembly

B Sulphuric acid

H_2SO_4, Conc. (Sp. gr. 1. 84)

C Hydrochloric acid, 0.1 N

See determination of alkalinity in water.

D Digestion catalyst

Grind together 20 g copper sulphate, 3 g mercuric oxide (HgO) and 1 g of selenium powder. Mix thoroughly 1 part of this mixture with 20 parts of sodium sulphate (Na_2SO_4) or potassium sulphate (K_2SO_4)

E Sodium hydroxide, 40%

Dissolve 40 g of NaOH in 100 ml of distilled water.

F Boric acid + mixed indicator. See determination of ammonia in water.

CHLORIDES

Most of the chlorides are soluble in water and can be determined directly in soil solution by titrating them with silver nitrate using potossium chromate as indicator.

ORGANIC MATTER

Reagents

A Potassium dichromate, 1 N

Dissolve 49.040 g of $K_2Cr_2O_7$ (A.R.) in distilled water to prepare 1 litre of solution,

B Ferrous ammonium sulphate, 1 N.

Dissolve 393.130 g of Fe $(NH_4)_2(S0_4)_2 6H_2O$ in distilled water adding simultaneously 15 ml of concentrated H_2SO_4 to prevent the hydrolysis of the salt. Dilute to 1 litre by distilled water and standardize by titration with $K_2Cr_2O_7$ solution.

C Sulphuric acid, concentrated.

Take H_2S0_4 (sp. gr. 1.84) and add 1.25 g silver sulphate to each 100 ml of acid.

D Diphenylamine indicator

Dissolve 0.5 g diphenylamine in a mixture of 100 ml conc. H_2SO_4, and 20 distilled water.

E Phosphoric acid

H_3PO_4, 85% (sp. gr. 1.71)

Procedure

Weigh suitable quantity of 0.5 mm perfectly dried soil not exceeding 10 g (containing about 10-25 mg carbon) and transfer it to a dried 500 ml conical flask. Add 10 ml of 1 N.$K_2Cr_2O_7$ and 20 ml concentrated sulphuric acid having silver sulphate dissolved in it, and mix by gentle swirling. Allow to stand the flask for about 30 minutes, and after when reaction is over, dilute the contents by 200 ml of distilled water. Add 10 ml of phosphoric acid and I ml diphenylamine indicator. The colour will change to bluish purple. Titrate the contents with ferrous ammonium sulphate, carefully, until the blue colour changes to brilliant green. The end point is very sharp in this titration. If more than 8 ml of the 10 ml added $K_2Cr_2O_7$ is consumed, (titration value < 2 ml) repeat the procedure with less quantity of the soil. If before the titration only, the colour becomes green, repeat with smaller quantity of soil.

Calculation

$$\% \text{ Carbon} = \frac{V_1 - V_2}{W} \times 0.003 \times 100$$

$$\% \text{ Organic matter} = \%C \times 1.724$$

where,

V_1 = Volume of $K_2Cr_2O_7$ (10 ml)

V_2= Volume of ferrous ammonium sulphate

W = Weight of the soil taken.

Depending on the soil, only 60-90% of the total organic matter is recovered in this method. No recovery factor has been taken into account in the formula given. Therefore the results should be accompanied by with a mention "by Walkley and Black method".

CALCIUM CARBONATE

Rapid Titration Method

Reagents

A. Hydrochloric acid, 1 N

Dilute 175 ml of concentrated HC1 to 2 litres. Standardization is not required in this case.

B. Sodium hydroxide, I N

Dissolve slightly more than 40 g of NaOH in I litre of distilled water and standardize with a standard acid. Dilute to the exact strength.

C. Brom thymol blue

Weigh 0.1 g of brom thymol blue and dissolve in 1.6 ml of 0.1 N NaOH by grinding in an agate mortar, and dilute to 250 ml.

Procedure

Take exactly 5 g of dried soil and trasfer to 150 ml beaker.

Add exactly 100 ml of I N HCI preferably with a pipette, and cover with a watch glass. Allow to stand for at least 1 hour with vigorous stirring intermittently. Allow to settle finally, and remove 20 ml of supernatent with the help of a pipette in a conical flask. Add 6-8 drops of brom thymol blue indicator and titrate with I N NaOH, at the end point the colour with change to blue.

Carry out a blank determination by using the same chemicals

but without soil

Calculation

$$\%CaCO_3 = (B - T) \times 5$$

where,

B = reading with blank

T = reading with soil

8

Key for Identification of Freshwater Organisms

IDENTIFICATION OF FRESH WATER COMMON FOOD FISHES

Identification of fresh water fishes was attempted by several authors. Besides these keys were prepared by several Ichthyologists to identify food fishes occurring in their own locality. Based on the information provided by the authors a practical key for identifying food fishes is suggested by the authors which helps in quick identification and for the benefit of the students.

Body Not Eel-like

A. Head and body scaled

(a)	Body laterally compressed and mouth not produced into a beak.	1. **Notopterus**
(*b*)	Body not laterally compressed and mouth produced into a beak.	2. **Xenentodon (Belone)**

B. Head scaleless and body scaled

(*a*)	Abdomen serrated	
(*i*)	Upper jaw not projecting and eyes with free adipose lids.	3. **Clupea**
(*ii*)	Upper jaw prominent and eyes covered by skin.	4. **Engraulis**
(*b*)	Abdomen edge cutting	5. **Chelu**
(*c*)	Abdomen neither cutting or serrated.	

(*i*) Air bladder enclosed in a bony capsule	
1. An erectile spine near orbit	
(*x*) Caudal forked.	6. **Botia**
(*y*) Caudal entire	7. **Lepidocephali ohthys**
2. No erectile spine near orbit.	8. **Nemachilus**
(*ii*) Air bladder not enclosed in bony	
1. Dorsal fin commencing nearly opposite the ventrals, Anal short.	
(*x*) Chin with a suctorial disc.	9. **Garra (Discognathus)**
(*y*) Lips thick and of two jaws continuous at the angle.	10. **Labeo**
(*z*) A small tubercle above symphysis of lower jaw.	11. **Cirrhinus**
(*z'*) Lower lip with a continuous transverse fold.	12. **Catla**
(*z''*) Lips of usual type	13. **(Barbus) Puntius**
2. Dorsal fin commencing posterior to ventral but not extending to or above anal.	
(*a*) Barbels present.	14. **(nuria)** ***Esomus***
(*b*) Barbels absent.	
(*x*) Mouth oblique, lower jaw with central and two lateral prominences and with a *black lateral stripe*.	15. **Rasbora**
(*y*) Mouth inferior and lower jaw with a sharp cutting edge.	16. **Aspidoparia**
3. Dorsel fin commencing above the interspace between the ventral and	

anal and generally extending above anal.

(*a*) Dorsal fin with osseous serrated ray. 17. **Osteobrama (Rohtee)**

(*b*) Dorsal fin without osseous ray. 18. **Barilius**

C. Head and body both scaleless

(*a*) An adipose dorsal fin absent.

1. Barbels eight.

(x) Dorsal fin long. 19. **Clarias**

(y) Dorsal fin short. 20. **Heteropneustes (Saccobra-nchus)**

2. Barbels less than eight. Cleft of mouth extending beyond the eyes. 21. **Wallago**

(y) Cleft of mouth not extending to eyes. 22. **Ompok (Cal -lichrous)**

(b) Anadipose dorsal present

1. Dorsal fin absent. 23. **Ailla**

2. Dorsal fin present.

(i) Barbels 2-4

(x) Anals 31-34 rays. 24. **Pangasius**

(y) Anals 40-46 rays. 25. **Silonia**

(ii) Barbels 6-8.

1. Adipose dorsal small.

(x) Fins emarginate. 26. **Bagarius**

(y) Fins not emarginate.

1. Eyes subcutaneous. 27. **Gagata**

2. Eyes not subcutaneous.

(i) Spines of dorsal and pectoral fins strong. 28. **Rita**

(ii) Spines of fins not strong.

(x) Anal fin of 10-12 rays. 29. **Nangra**

(y) Anal fins of 29-57 rays,

(z) Nares patent anal fin 29-36 rays. 30. **Clupisoma (Pseudeutropius)**

(z) Nares usual.

1. Anal fin 44-51 rays. 31. **Eutropiichthys**
2. Adipose dorsal long. 32. **Mystus**

II. Body eel-like 33. **Amphipnous**

Fins with Some Rays Spinous

Body not eel-like and ventral fins present

(a) Ventral fins confluent. 34. **Glossogobius (Gobius)**

(*b*) Ventral fins produced into a pair of filaments. 35. **Colisa (Trichogaster)**

(c) Ventral fins usual.

(a) Body clindrical.

1. A single spinous dorsal fin. 36. **Channa (Ophiocephalus)**
2. A small spinous and another, soft dorsal fin. 37. **Rhinomugil (Mugil)**

(b) Body laterally compressed.

1. Caudal forked. 38. **Ambasis**
2. Caudal rounded and body striped. 39. **Nandus**

Body Eel-like and Ventral Fins Absent

(a) Preorbital spine present no ocelli present. 40. **Mastocembelus**

(*b*) Preorbital spine absent and soft dorsal with ocelli. 41. **Macrognathus (Rhynchobdella)**

AGONISTIC BEHAVIOUR IN FISH

Background

Several studies of agonistic behaviour in various species of fish have been reported. Considerable data is available on the agonistic behaviour of the three-spined stickle back. Gasterosteus aculeatus. (Tinbergen, 1951). Atlantic salmon, Salmon, Salmo salar L., (Keenleyside and Yamamoto, 1962), and Siamese fighting fish, Betta splendens, (Thompson, 1969). A comparison of agonistic behaviour in several species of fish was reported by McDonald and Heimstra (1965). Pairs of fish, each representing one of the seven species used, were placed in separate aquaria and observed for five minutes on each of five consecutive days. The bluegill sunfish. Lepomis macrochirus, was found to exhibit the highest frequency of attacks towards conspecifics.

Reproduction, feeding, and aggressive behaviours of longer sunfish, *Lepomis megalotis* in nature and in the laboratory have been observed by Huck and Gunning (1967). Confinement of two long-ear sun-fish in a five-gallon aquarium brought about an increase in agonistic behaviour. Under this condition, dominance was always established by one fish within a matter of hours. Overtly aggressive behaviour continued, with the dominant fish nipping the subordinate until death of the latter resulted. Laboratory observed aggression of long-ears was viewed as an artifact of the laboratory situation, since the physically aggressive behaviour observed in the laboratory was not evident in the natural environment.

The inability of the subordinate individual to escape from the dominant one, when both are held in a small aquarium, must be considered as an important factor contributing to the higher frequency of laboratory observed aggression. In a natural setting, the sub-ordinate fish may terminate an encounter by swimming away.

This study is designed to explore the role played by the local size of the resident and intruder as a determinant of agonistic behaviour. By allowing at least a 7-day residence period prior to introduction of an intruder, it may be possible to specify a priori who will be dominant and who will be submissive in as much as length of residence in the living space may influence the degree to which a sunfish will behave aggressively to an "intruder".

With the help of your instructors, conduct a team research project related to agonistic behaviour by bluegill sunfish (Lepomis macrochirus). Other species of panfish such as pumpkinseed sunfish, and green sunfish may be available for expansion of your study.

Objectives

Findings by preparing a written report, abstract and oral report in appropriate scientific style.

Secondary objectives include practice in making (1) observations, 9, (2) formulating hypotheses, (3) designing an experiment based upon one of your hypotheses, (4) conducting experiments, (5) gathering and recording data, (6) analyzing and interpreting your data (7) graphing and/or tabulating data, and (8) drawing appropriate conclusions.

Suggested Work Plan

Familiarize yourself with the external anatomy of a fish.

Experimental Design

This experiment is designed as a factorial manipulation of resident and intruder size, combined with analyses of aggressive behaviour as a function of both size and time.

A 2 × 2 × 6 factorial experimental design consisting of one within and two between variables is summarized in Table 1.

Size of the resident and intruder represent the orthogonally (rectangular) arranged independent variables. Time, the one within variable, is represented in terms of successive 5 minute intervals of the 30 minute observation period. Frequencies of identified and defined agonistic behaviours exhibited by resident to intruding conspecifies serve as the dependent variation.

Resident Size

	Large	Small
1.		

....

2. Time Periods

....

3.

.... Aquarium 3

4. Aquarium I

....

5.

....

6.

....

Aquarium 2 Aquarium 4

Familiarize yourself with the list and description of agonistic behaviours found in Appendix I. Note that these were written in respect to behaviour of a resident (R), but can also apply to an intruder (I).

Sit as far back from the aquarium as possible and remain motionless as possible. If you move the fish might watch you instead of the intruder. One student in each group should record data, one should call out time periods and one function as the official observer. Time may begin when an intruder (opposite size of resident) is placed in the aquarium. At the end of 30 minutes remove the intruder and return to the stock tank.

Graph number of Ambivalent acts in Figure 1
Graph number of Preparatory acts in Figure 2
Graph number of Overtly Aggressive acts in Figure 3
Graph total number of acts in Figure 4

Each student research group should determine the total number of agonistic behaviours for both residents and intruders in aquariums 2 and 3. Therefore, only one-half of the experimental design will be looked at. Aquarium number Residents Intruders

1	2	3	4
Large	Large	Small	Small
Large	Small	Large	Small

Formulate hypotheses based upon your results, analysis and interpretation. Record on the laboratory report form.

Design a subsequent experiment based upon your hypothesis. Draw the design in graphic form on the back of the page titled Laboratory Report. What follows is a potential and logical extension of your first experiment. However, do not hesitate to use your imagination and ingenuity in your design keeping in mind the limits of your facility and apparatus.

SCHOOLING AND AGGREGATION IN FISH BACKGROUND

Many species of fish tend to undergo aggregation at various stages of their life history and under various types of environmental conditions. An aggregation is often called a school and defined as "an aggregation formed when one fish reacts to one or more other fish by staying near them". The principal characteristics of the organization of fish schools are that the individuals stay together, tend to head in the same direction, maintain even spacing, and the activities of individuals tend to be synchronized. Though not well-understood the mechanism for the cohesion and synchronization of a school is considered to be a "following reaction" common to flocks of birds.

Schooling usually develops gradually in nature and may be purely a result of maturation. Furthermore, a number of studies have reported the rapid appearance of normal schooling when fish raised in isolation are placed with conspecifies, i.e., belonging to the same species. Recent research indicates that isolates placed together after 20 days of age school promptly, whereas isolates placed together before this age do not.

However, continual readjustment of position and reorientation suggested schooling of isolated was slightly abnormal.

It has been suggested that one of the possible explanations of the adaptive advantages of schooling is that it facilitates learning. Better learning performance occurred in grouped fish than isolated fish and many authors assume that schooling plays a substantial role both in explaining the adaptive value and its development.

The safety of the group seems to be another advantage of the schooling habit, since the reaction to the approach of predators varies among species. Herring school closely on the approach of danger, and anchovies under attack by predators crowd together-forming dense, compact balls. Other advantages, often cited, for schooling include facilitation of finding food and aids in reproduction. When it is time to reproduce, there is no courtship behaviour, no male selection, and males and females simply shed eggs and sperms in almost countless numbers, enhancing the probability of successful fertilization. Another advantage suggested for schooling is that it provides a more effective way of moving through water. The effort of each fish ın the school may be lessened because it can utilize the turbulence created by surrounding fish.

Breder (1959) describes a polytipic (i.e., in this case two different species) aggregation (schooling) between the minnows Notemigonus crysoleucas (Mitchell) and Erimyzon sucetta (Lacepede) whereas Nursall and Pinsent report a grouping of potential but immature predators, yellow perch (Perca fluviatilis) and prey, the spottail shiner, (Norropis) hudsonius (Clinton).

Size ranges of the later aggregations were nearly the same ranging from 30 to 110 mm. Each combined aggregation was observed to take a definite position in water 1.3 m deep with the upper most fish 15-20 cm below surface and the lowermost fish 30-50 cm above the bottom. Distance of the aggregation from the bottom is greater in deeper water. Likewise, there is a size/depth stratification with smaller fish tending to stay near the top and larger individuals more common near the bottom. The individual size and position relationship is more marked for the spottail shiner than the yellow perch. As perch increase in size over the growing season to 110 mm they drop out of the aggregation. There is a tendency towards species stratification with the spottail shiners being more abundant in the upper aggregation layers and perch concentrating in the lower. This stratification is definite, but not fixed. `Shiners and perch can be found at all levels with a broad zone of intermingling of the species.

With the help of your instructor conduct an extended research project related to species, size and behaviour in schooling fishes. Note that only part of the research project is provided and you are expected to extend the research at time permits.

SUGGESTED WORK PLAN

Observation and Hypothesis

Observe and record the behaviours of the available fish. Pay particular attention to the aggregation tendencies of the various species.

In addition to others you should ask the following questions. Is aggressive behaviour obvious among fish of the same species? Between the available species? Do the fish aggregate in a single-species group? What is the relationship between density and spacing of schooling and the preferred habitat within the aquarium? Does aggregation or schooling behaviour breakdown when feeding or under stressful conditions?

Experiment I

Conduct the experiment according to the following design. Note that the experimental aquarium is divided into three equal areas. Add aquarium water and either or ten fish of one species to a one gallon jar with a screw top lid. Repeat this process for a second species. Place the individual jars in the far right and far left areas of the experimental aquarium. Net a Test Fish or Isolator of Species I from the stock tank and gently release it in the center of the aquarium making sure that it is lot directed toward one or the other of jars.

Record time in minutes the Isolator spends in each of the three areas and the number of times the Isolatar moves from one area to another:

Remove the Isolator, reverse the positions of the jars, replace the Isolator and repeat the data collection.

Repeat the process as many times as necessary to collect statistically reliable data.

Now repeat all of the above with an Isolator from Species II.

Record behaviuors of fish in the jars and Isolators as appropriate.

Experiment 2

This investigation attempts to determine the effect of fish in a group on the behaviour. of an isolated fish. The variable to be tested is different size of groups.

Place two jars in opposite sides of the aquarium and add two Species I fish to one and six Species I fish to the second.

Release a conspecific fish at the midline of the aquarium. During a 15 minute time period determine amount of time spent in each of the three areas and the number of times the Isolator moves from one area to the next.

Remove the Isolator and use others as necessary for a statistically reliable sample.

Following repeated trials reverse the positions of the jars, replace the priginal isolators and record your data.

When finished place equal numbers of fish in each jar and repeat as above.

What will happen if instead of 6 fish as above, 12 fish are used?

FOLLOW-UP

Formulate several written hypotheses based upon your newly acquired knowledge. Design subsequent experiments based upon your hypotheses. Draw the experimental design in graphic form on the back of the page titled laboratory report and discuss it with your instructor.

Experiment 3

Test your newly-formulated hypotheses by conducting your appropriately designed experiments. Be sure to replicate necessary and record all data for your appendix.

Prepare additional tables and figures as necessary.

FOLLOW-UP

Begin to prepare a scientific paper and abstract according to acceptable style and format. Slides and other visuals should be planned and prepared for your oral presentation.

9

Study of Animal Population

THE SAMPLING PROGRAMME

The Number of Samples Per Habitat Unit (e.g. plant)

There are two aspects, firstly whether different regions of the unit need to be sampled separately and secondly the number of samples within each unit or subunit (if these are necessary) that should be taken for maximum efficiency. Although the habitat unit could, for example, be the fleece of a sheep, a bag of grain or a rock in a stream, for convenience the word plant will, in general, be used in its place in the discussion below.

Subdivision of the Habitat

If the distribution of the population throughout the habitat is biased towards certain subdivisions, but the samples are taken randomly, what aptly term *systematic errors* will arise. This can be overcome either by sampling so that the differential number of samples from each subdivision reproduces in the samples the gradient in the habitat, or by regarding each part separately and correcting at the end. The amount of subdivision of the plant that various workers have found necessary varies greatly. On apple the eggs, larvae and pupae of the tortricid moth, *Archips argyrospilus,* were found for most of the year to be randomly distributed over the tree so that only one level (the lower for ease) needed to be sampled. In contrast on the same trees and in the same. years the immature stages of two other moths showed marked differences between levels at all seasons. With the spruce budworm *(Choriston-eura fumiferana),* found that there were 'substantial and significant differences from one crown level to another' and that there was a tendency for eggs and larvae to be more abundant at the top levels, but there was no significant difference associated with different sides of the same tree. A similar variation with height was found with the eggs of the larch sawfly *(Pristiphora erichsonii),* although here it was concluded that in view of the cost and mechanical difficulties of stratified sampling at different heights a reasonable index of the population would be obtained by sampling the mid-crown only. Studying all the organisms on aspen *(Populus tremuloides)* found it is necessary to sample at three different levels of the crown and even with field crops height often needs to be considered: recommended that potato aphids be estimated by picking three leaves - lower, middle and upper - from each plant and when estimating the population of the european corn borer *(Ostrinia nubilalis),* showed that the lower and the upper halves of the maize stem needed to be considered separately, the former containing the majority of the larvae. The distribution of the eggs of various cabbage-feeding Lepidoptera was found by to depend on the age of the plant.

Aspect is sometimes important; in Nova Scotia in the early part of the season the codling moth lays mostly on the south-east of the apple trees, but later this bias disappears. Aspect has also been found to influence the distribution on citrus of the long-tailed mealy bug, *Pseudococcus adonidum*, the eggs of the oak leafroller moth, *Archips*, and of three species of mite, each of which was most prevalent on a different side, but not that of the pine beetle, *Dendroctonus*. Variations in the spatial distribution of similar species in the same habitat, which complicates a sampling programme designed to record both, has also been recorded for two potato aphids by Helson. Some insects are distributed without bias on either side of the mid vein of leaves, so that they may be conveniently subsampled record that when estimating populations of sheep keds, the fleece of only one side need be sampled.

Occasionally it may be found that such a large and constant proportion of the population occur on a part of the plant that sampling may be restricted to this: showed that in Minnesota 84 %

of the eggs of the spruce budworm *(Choristoneura fumiferana)* are laid on the tips of the branches and if sampling is confined to these, rather than entire branches, sampling time may be reduced by up to 40%.

The taking of a certain number of samples randomly within a site which is itself selected randomly from within a larger area, e.g. a field, is often referred to *as nested sampling,* and may be on two, three or more levels.

The Number of Samples Per Subdivision

To determine the optimum number of samples per plant (or part of it) (n), the variance of within-plant samples $\left(s_s^2\right)$ must be compared with the variance of the between-plant samples $\left(s_p^2\right)$ and set against the cost of sampling within the same plant (c_s) or of moving to another plant and sampling within it (c_p):

$$n_s = \sqrt{\frac{s_s^2}{s_p^2} \times \frac{C_p}{C_s}}$$

If the interplant variance (s_p^2) is the major source of variance and unless the cost of moving from plant to plant is very high n will be in the order of one or less (which means one in practice). Interplant variance has been found to be much greater than within-plant variance for the spruce sawfly *(Diprion hercyniae)*, the lodgepole needle miner *(Recurvaria starki)*, the cabbage aphid *(Brevicoryne brassicae)*, the spruce budworm *(Choristoneura fumiferana)*, the winter moth *(Operophtera brumata)*, the diamondback moth *(Plutella maculipennis)*, the cabbage butterfly *(Pieris rapae)*, the Western pine beetle *(Dendroctonus brevicomis)*, the pine chermid *(Pineus pinifoliae,)* and the spider mite *(Panonychus ulmi,)*. In most of these examples the within-plant variance was small so that only one sample was taken per plant or per stratum of that plant, although of course when this is done, within and between tree variances cannot be separated. With some apple insects the within-tree variance (s_s^2) becomes larger, especially at certain seasons, and then as many as seven samples may be taken from a single tree.

Often a considerable saving in cost without loss of accuracy in the estimation of the population, but with loss of information on the sampling error, may be obtained by taking randomly a number of subsamples which are bulked before sorting and counting. This is especially true where the extraction process is complex as with soil samples; bulked four random, 3-in row samples of young oat plants and soil to make a single 1-ft row sample that was then washed and the eggs of the frit fly *(Oscinella frit)* extracted. Such a process gave a mean as accurate as that obtained by washing all the 3-in samples separately (greater cost). Sampled the eggs of a tortricid moth, *Archips argyrospilus,* on apple by bulking 25 cluster samples.

The sampling unit, its selection, size and shape

The criteria for the sample unit are broadly (Morris, 1955):

(1) It must be such that all units of the universe have an equal chance of selection.

(2) It must have stability (or if not its changes should be easily and continuously measured -as with the number of shoots in a cereal crop).

(3) The *proportion* of the insect population using the sample unit as a habitat must remain constant.

(4) The sampling unit must lend itself to conversion to unit areas.

(5) The sampling unit must be easily delineated in the field.

(6) The sampling unit should be of such a size as to provide a reasonable balance between the variance and the cost.

(7) The sampling unit must not be too small in relation to the animal's size as this will increase edge effect errors.

(8) The sampling unit for mobile animals should approximate to the average ambit of an individual. This 'condition' is particularly significant in studies on dispersion involving contiguous sampling units has suggested that a test of the appropriate size would be provided by

several series of counts of animals in contiguous quadrats conforming to a Poisson series.

A sampling unit defined in relation to the animal's ambit or territory (e.g. gallery of a bark-beetle), will give different information from one defined in terms of the habitat. This re-emphasizes the need to be very clear as to objectives and hypothesis before commencing a sampling programme.

To compare various sampling units in respect to variance and cost it is generally convenient to keep one or other constant. The same method of sampling must, of course, be used throughout. From preliminary sampling the variances of each of the different units (s_u^2) can be calculated; these should then be computed to a common basis, which is often conveniently the size of the smallest unit. For example if the smallest unit is 1 ft of row, then the variance of 2 ft row unit will be divided by 2 and those of 4 ft by 4. The costs will similarly be reduced to a common basis (C_u). Then the relative net cost for the same precision for each unit will be proportional to:

$$C_u s_u^2$$

where C_u =cost. per unit on a common basis and s_u^2 =variance per unit on a common basis. Alternatively the relative net precision of each will be proportional to $1/C_u s_u^2$. The higher this value, the greater the precision for the same cost.

A full treatment of the methods of selecting the optimum size sampling unit is given in Cochran and other textbooks, but as population density, and hence variance, is always fluctuating, too much stress should not be placed on a precise determination of optimum size of the sampling unit.

Even with soil animals, where such a procedure might be of most value, showed that although 4- and 6-in diameter cores are equally efficient at low densities, at high densities the comparative efficiency of the 6-in sample falls off. With insects on plants the nature of the plant usually restricts the possible sizes to, for example, half leaf, single leaf, or shoot.

The shape of the sampling unit when this is of the quadrat type, rather than a biological unit, is theoretically of importance because of the bias introduced by edge effects. These are minimal with circles, maximal with squares and rectangles and intermediate with hexagonals, because they are proportional to the ratio of sample unit boundary length to sampling unit area. If the total habitat is to be divided into numbered sampling units (for random number selection), then circular units are impractical because of the gaps between and it is doubtful if the reduction of error from the use of hexagons normally justifies the difficulties of lay-out. Clearly the larger the sampling unit, proportionally the less the boundary edge effect. The size of the organism will also influence this effect: the larger it is in relation to the sample size, the greater the chance of an individual lying across a boundary. This problem has been investigated for subcortical insects where the damage to the edge individuals by the punch, and the curved nature of the sampled substrate pose special problems. In general edge effects can be minimized by a convention (e.g. of the animals crossing the boundaries only those on the top and left-hand boundaries are counted).

The number of samples

The total number of samples depends on the degree of precision required. This may be expressed either in terms of achieving a standard error of a predetermined size, or in probability terms, getting confidence limits of a predetermined half-width, a percentage of the mean.

For many purposes a standard error of 5 % of the mean is satisfactory. Within a homogeneous habitat the number of samples (n) required is given by:

$$n = \left(\frac{s}{E\bar{x}} \right)^2$$

where s = standard deviation, x = mean and $\boldsymbol{E}$ = is the predetermined standard error as a decimal of the mean (i.e. normally 0.05). This expression compares the standard deviation (s) of the observations with the standard error ($E\bar{x}$) acceptable for the contrasts we need to make; it will be noted from this equation that in any given situation the value of the standard error will change with the square root of the number of samples: thus a large increase in n is necessary to bring about a small improvement in s.

Where sampling is necessary at two levels, e.g. a number of clusters per tree, the number of units (n_1) that need to be sampled at the higher level, e.g. trees is given by:

$$n_t = \frac{\left(s_S^2 / n_S\right) + S_p{}^2}{(\bar{x} \times E)^2}$$

where n_s = the number of samples within the habitat unit (calculated as above), $s_S{}^2$ = variance within the habitat unit, $s_p{}^2$ = variance between the habitat unit (= interplant variance), x = mean per sample (calculated from the transformed data and given in this form and E as above.

Rojas (1964) has shown that if the dispersion of the population has been found to be well described by the negative binomial the desired number of samples is given by:

$$N = \frac{1/\bar{x} + 1/k}{E^2}$$

where k =the dispersion parameter of the negative binomial (see below).

When the confidence limits are used as the predetermined standard, the required half-width is usually set at 10'% of the mean. The general formula * then becomes:

$$n' = \left(\frac{ts}{Dx}\right)^2$$

where t = 'Students t' of standard statistical tables, it depends on the number of samples and approximates to 2 for more than 10 samples at the 5°, level, and D = the predetermined half-width of the confidence limits as a decimal (usually 0.1). It will be seen that normally this gives a similar estimate to equation 2.7, provided the values of Eand D are adjusted to accord with their meanings. The procedure is perforce somewhat approximate, depending on the preliminary estimates of the mean and standard deviation, and the inclusion of t does perhaps give it a slightly bogus air of precision! Additionally there is normally a biological approximation, for as population characters change with time, so will the optimal number of samples.

Another type of sampling programme concerns the measurement of the frequency of occurrence of a particular organism or event, for example the frequency of occurrence of galls on a leaf or of a certain genotype in the population. Before an estimate can be made of the total number of samples required, an approximate value of the probability of occurrence must be obtained. For example, if it is found in a preliminary survey that 25% of the leaves of oak trees bear galls the probability is 0.25. The number of samples (N) is given by:

$$N = \frac{t^2 pq}{D^2}$$

where p = the probability of occurrence (i.e. 0.25 in the above example), $q = 1 - p$; t and D are as above.

If it is found that the leaves (or other units) are distributed differently in the different parts of the habitat, they should be sampled in proportion to the variances. For example, Henson found from an analysis of variance of the distribution of the leaf-bunches of aspen that the level of the crown from which the leaves had been drawn caused a significant variation and when this variance was. portioned into levels the values were: lower 112993, middle 68012, upper 39436. Therefore leaf-bunches were sampled in the ratio of 3 : 2: 1 from these three levels of the crown.

THE PATTERN OF SAMPLING

Once again it is important to consider the object of the programme carefully. If the aim is to obtain estimates of the mean density for use in, for example, life-tables, then it is desirable to minimize variance. But if the dispersion

(= distribution = pattern) of the animal is of prime interest then there is no virtue in a small variance.

In order to obtain an unbiased estimate of the population the sampling data should be collected at *random*, that is so that every sampling unit in the universe has an equal chance of selection. In the simplest form -the *unrestricted random sample -the* samples are selected by the use of random numbers from the whole area (universe) being studied (random number tables are in many statistical works, or the last two figures in the columns of numbers in most telephone books provide a substitute). The position of the sample site is selected on the basis of two random numbers giving the distances along two co-ordinates, the point of intersection is taken as the centre or a specified corner of the sample. If the size of the sample is large compared with the total area then the area should be divided in plots which will be numbered and selected using a single random number. Such a method eliminates any personal choice by the worker whose bias in selecting sampling sites may lead to large errors.

However, just because it is absolutely random this method is not very efficient for minimizing the variance, since the majority of the samples may turn out to come from one area of the field. The method of *stratified random* sampling is therefore to be preferred for most ecological work; here the area is divided up into a number of equal sized subdivisions or strata and one sample is randomly selected from each strata. Alternatively if the strata are unequal in size the number of units taken in each part is proportional to the size of the part; this is referred to as self-weighting. Such an approach maximizes the accuracy of the estimate of the population, but an exact estimate of sampling error can only be obtained if additional samples are taken from one (or two) strata. The taking of one sample randomly and the other a fixed distance from it has been recommended by Hughes as a method of mapping aggregations. The fixed distance must be less than the diameter of the aggregations and assumes these are circular; the standard error cannot be calculated. However the method has been found useful for soil and benthnic faunas.

When the habitat is stratified biological knowledge can often be used to eliminate strata in which few insects would be found. Such a restricted universe will give a greater level of precision for the calculation of a mean than an unrestricted and completely random sample with a wide variance. Found with a pine sawfly:

satisfactory estimates of the pupae were only obtained if sampling was limited to the areas around the bases of the trees, the variance of completely random sampling throughout the whole forest was too great, as many areas were included that were unsuitable pupation sites.

The other approach is the *systematic sample,* taken at a fixed interval in space (or time). In general such data cannot be analysed statistically, but Milne has shown that if the *centric systematic area-sample* is analysed as if it were a random sample, the resulting statistics are 'at least as good, if not rather better', than those obtained from random sampling. The centric systematic sample is the one drawn from the exact centre of each area or stratum and its theoretical weakness is that it might coincide with some unsuspected systematic distribution pattern. As Milne points out, the biologist should, and probably would, always watch for any systematic pattern, either disclosing itself as the samples are recorded on the sampling plan or apparent from other knowledge. Such a sampling programme may be carried out more quickly than the random method and so has a distinct advantage from the aspect of cost.

An example of an unbiased systematic method is given by Anscombe. All the units (e.g. leaves) are counted systematically (e.g. from top to bottom and each stem in turn), then every time a certain number (say 50) is reached that unit is sampled and the numbering is commenced again from 1; only one allocation of a random number is needed and that is the number (say somewhere between 1 and 20) allotted to the first unit.

Biologists often use methods for random sampling that are less precise than the use of random numbers, such as throwing a stick or quadrant or the haphazard selection of sites. Such methods are not strictly random; their most serious objection is that they allow the intrusion of a personal bias, quite frequently marginal areas tend to be under sampled.

It may be worthwhile doing an extensive trial comparing a simple haphazard method with a fully randomized or systematic one, especially if the cost of the latter is high when compared with the former. Spiller found that scale insects on citrus leaves could be satisfactorily sampled by walking round the tree, clockwise and then anticlockwise, with the eyes shut and picking leaves haphazardly. For assessing the level of red bollworm eggs (*Diparopsis castanea*) to determine the application of control measures Tunstall & Matthews recommended two diagonal traverses across the field counting the eggs at regular intervals.

Bias may intrude due to causes other than personal selection by the worker: grains of wheat that contain the older larvae or pupae of the grain weevil *(Sitophilus granarius)* are lighter than uninfested grains. The most widespread method of sampling is to spread the grains over the bottom of a glass dish and then scoop up samples of a certain volume; as the lighter infected grains tend to be at the top this method can easily overestimate the population of these stages. In contrast, for the earlier larval instars before they have appreciably altered the weight of the grain such a simple method gives reliable results; it was undoubtedly this difference that led to Krause & Pedersen stressing the need for samples to contain a relatively high proportion of the same stage if good replication was to be obtained.

The timing of sampling

The seasonal timing of sampling will be determined by the life-cycle of the insect. In extensive work when only a single stage is being sampled, it is obviously most important that this operation should coincide with peak numbers. This can sometimes be determined by phenological considerations, but the possibility of a control population in an outdoor cage (Harcourt) to act as an indicator should be borne in mind. The faster the development rate, the more critical the timing. With intensive studies that are designed to provide a life-table regular sampling will be needed throughout the season.

It is not always realized that the time of day at which the samples are taken may also considerably affect them. The diurnal rhythms of the insects may cause them to move from one part of the habitat to another as Dempster found with the Moroccan locust *(Dociostaurus maroccanus)* (Fig. 9.1).

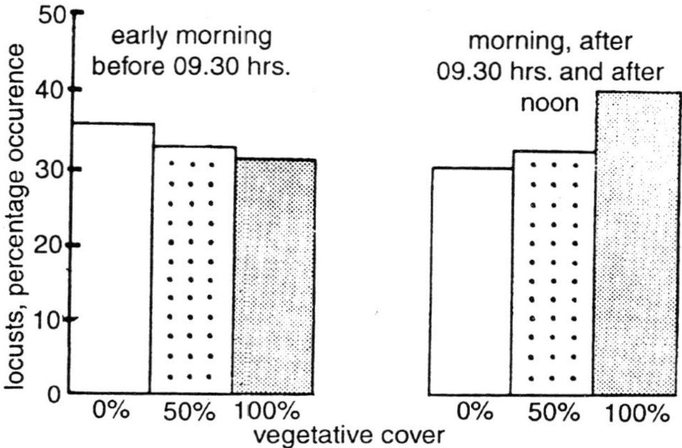

Fig. 9.1: ***The variation in the distribution of adults of the Moroccan locust Dociostaurus maroccanus at different times of the day; the histograms show the relative numbers on bare ground and areas with moderate and dense vegetation.***

Many grassland insects move up and down the vegetation not only in response to weather changes, but also at certain times of the day or night and during the day quite a proportion of active insects may be airborne (cf. the observations of Southwood, Jepson & Van Emden on the numbers of adults of the frit fly (*Oscinella frit*) on oats). There is a marked periodicity of host-seeking behaviour in many blood-sucking invertebrates. The ecologist may find that some of his sampling problems can be overcome, or at least additional information gained, if he works at night or at dusk and dawn, rather than during conventional working hours.

SEQUENTIAL SAMPLING

In this type of sampling the total number of samples taken is variable and depends on whether or not the results so far obtained give a definite answer to the question posed about the frequency of occurrence of an event (i.e. abundance of an insect). It is of particular value for the assessment of pest density in relation to control measures, when these are applied only if the pest density has reached a certain level. Extensive preliminary work is necessary to establish the type of distribution (the relation of the mean and the variance) and the density levels that are permissible and those that we associated with extensive damage. From such data it would be possible to lay down a fixed number of samples that would enable one to be sure in all instances what was the population level; however, this would frequently lead to unnecessary sampling and a sequential plan usually allows one to stop sampling as soon as enough data have been gathered. For example, to take extremes, if 80 insects per shoot, or more, caused severe damage and no insects were found in the first 5 samples, common sense would tempt one not to continue for the additional, say 15, samples. Sequential sampling gives an exact measure (based on the known variance) so that with extremely high and extremely low populations very few samples need be taken and the expenditure of time and effort (cost) is minimal. It may also be used to obtain population estimates with a fixed level of precision.

As the distribution of most insect species can be fitted to the negative binomial, or to Iwao's Patchiness Regression, formulae will be given only for these types of distribution; of course the principles are the same with other distributions. The method is described by Wald, Goulden and Waters, Stark has applied a sequential plan to normalized data for a needle miner (Lepidoptera).

The first decision must be to fix the insect population levels related to the infestation classes; we wish to distinguish between the hypothesis (H_1) that there are e.g. 200 or more egg masses per branch, sufficient to cause heavy damage, and the hypothesis (H_0) that there are e.g. 100 or fewer egg masses per branch, insufficient to cause damage.

The second decision concerns the level of probability of incorrect assessment one is prepared to tolerate; there are two types of error:

ψ = the probability of accepting H_1 when H_0 is the true situation w

ω = the probability of accepting H_0 when H_1 is the true situation

Let us say that the same level is accepted for both, 1 false assessment in 20, i.e. a probability level of 0.05 for 0 and w. Different levels may be necessary because the costs of failing to apply control measures when an outbreak occurs, may be much higher than those arising from the application of the treatment that is proved not to have been required.

For an animal whose distribution corresponds to the negative binomial the following values need to be calculated; the common k having been found already and the means being fixed as above:

	Infestation level	
	H_0	H_1
Mean = kp	kp_0(e.g. 100)	kp_1 (e.g. 200)
$p = kP/k$	P_0	P_1
$q = 1 + p$	q_0	q_1
Variance = kpq	kp_0g_0	kp_1g_1

The next aim is to plot the two lines that mark the 'acceptance' and 'rejection' areas (these terms have come from quality control work where sequential sampling was originally developed).

The formulae for the lines are:

$$d_0 = \theta n + h_0$$
$$d_1 = \theta n + h_1$$

where d = cumulative number of insects, n = number of samples taken, θ = the slope of the lines given by

$$\theta = k\frac{\log(q_1 / q_0)}{\log(p_1 q_0 / p_0 q_1)}$$

and h_0 and h_1 the lower and upper intercepts given by

$$h_0 = \frac{\log[\omega / (1-\psi)]}{\log(p_1 q_0 / p_0 q_1)}$$

$$h_1 = \frac{\log[(1-\omega) / \psi]}{\log(p_1 q_0 / p_0 q_1)}$$

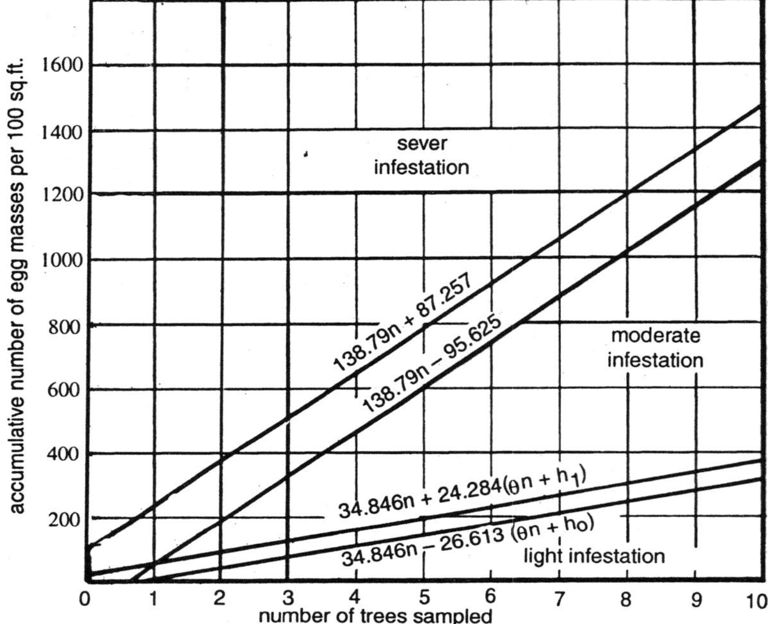

Figure 9.2 : *A sequential sampling chart with two sets of acceptance and rejection lines.*

Once these calculations are made a graph similar to Figure 10.2 can be drawn up, and it will be clear that occasionally (especially when the true population lies between the population levels chosen, i.e. 100 and 200 in this example) the cumulative total continues to lie in the uncertain zone and therefore an arbitrary upper limit must be set to the number of samples taken. This limit (x) will be decided on considerations of cost, and normally it will be laid down that if after x samples the result still lies in the uncertain zone treatment will be applied.

If Iwao's Patchiness Regression is used the lines can be best calculated for successive values of n (number of samples taken) by (Iwao):

$$d_1 = nx_c + t\sqrt{n\left[(\alpha + 1)x_c + (\beta - 1)x_c^2\right]}$$

$$d_0 = nx_c - t\sqrt{n\left[(\alpha - 1)x_c + (\beta - 1)x_c^2\right]}$$

where x_c = the critical density, α and β are the parameters from the patchiness regression and t is taken from statistical tables for the acceptable error levels for ψ (for d_0 equation) and ω (for the d_1 equation).

It is possible to draw up operating characteristic curves and average sample number curves which give a measure of the probability of accepting the two hypotheses and the average number of samples necessary, respectively, at different population levels.

The actual field work may be carried out based on the graph, or a sequential table may be prepared from it. This gives the uncertainty band for various numbers of samples; e.g. (after Morris,):

Sample tree	Moderate vs severe infestation uncertainty band (cumulative total of insects)
1	42-225
2	183-366
3	323-506
4	460-643
5	599-782
6	738-921

It is often considered desirable to set up three classes: severe, moderate and light, and then two pairs of lines are calculated.

A general account of the utility of sequential sampling plans for forest insects is given by Ives and, in addition to the papers already referred to, plans have been developed for the winter moth *(Operophtera brumata)*, the forest tent caterpiller *(Malacosoma disstria)*, the larch sawfly *(Pristiphora erichsonii)*, the red-pine sawfly *(Neodiprion nanulus)*, the coffee shield bug, *Antestiopis*, lodgepole needle miner *(Evagora milleri)*, the aphid, *Myzus persicae*, white grubs, *Pieris rapae*, wireworms, cotton fleahopper *(Pseudatomoscelis)*, cotton arthropods and others. It must, however, be remembered that these plans are based on particular sampling methods in a certain area with a given developmental stage and, as has been shown, the value of k and hence the validity of the plan may alter if any of these are changed.

PRESENCE OR ABSENCE SAMPLING

It is sometimes very costly to estimate the actual numbers in a sample, but presence or absence can be easily assessed. If the dispersion of such an animal in a particular habitat, e.g. forest, can be described by the negative binomial distribution with a known *k,* the probability of a particular mean population per tree (P_y) being reached can be estimated:

$$p_y = k\left[\left(\frac{1}{1-p'}\right)^{1/k} - 1\right]$$

where p' = probability of the animal being present in the sampling unit - determined on the basis of presence or absence sampling. The form of the relationship is shown in Figure 9.3. This method provides an alternative to sequential sampling as a basis for pest control decisions. It is particularly useful with highly clumped animals; the saving on counting time is greater, as is the sensitivity (Figure 9.3) and is reliable only if the critical density levels are related to values ofp' less than about 0.8; above this the uncertainty associated with the predictions is too great. Obviously the sampling unit can be varied to achieve this relationship but for it to retain its value for rapid assessment the new unit must be easily recognized. Furthermore any major changes in the value of k with density would invalidate this approach.

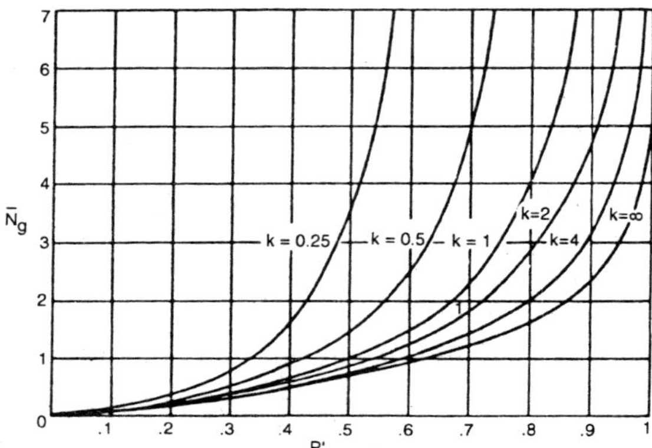

Figure 9.3 : ***Relationship between mean population density*** **(N_y)** ***and the probability of cell occupancy*** **(p')** ***for selected values of*** **k.**

SAMPLING A FAUNA

Emphasis has been placed on the problems involved in sampling the population of a single species and these are formidable enough. But with faunal surveys the problem of ensuring the detection of an adequate proportion of the species present must be considered. When sampling vegetation, Gleason pointed out as the area sampled increased the numbers of hitherto unrecorded species added decreased, a concept referred to as the species-area curve, and one that has been often used and disputed in plant ecology. The logarithmic series and log normal distributions) express the same basic assumptions another way.

One approach to this problem is therefore to take a number of preliminary samples and calculate the index of diversity based on the log series. If one can make an estimate of the total number of individuals in the fauna then the theoretical total number of species present can be arrived at from the index of diversity, and the total number of individuals that should be collected to find y % of species present can be determined. The value of y will, of course, depend on the aim of the study. Unfortunately, as pointed out later, this approach has a number of practical and theoretical limitations. The practical ones, stem from the mosaic nature of most habitats, and therefore, unless sampling is by a light trap or similar device that is independent of microhabitats, more new species are added by further samples than would be expected. The theoretical problems arise from the discovery that the log normal distribution provides a better description of the phenomenon than the simpler logarithmic series. An approach which makes no assumptions about the type of distribution is given by Good & Toulmin.

Sometimes the mosaic nature of the habitat can be utilized to facilitate the planning of an optimal sampling programme. This is especially true where the effects of pollution are being studied in streams, but might be extended to other situations where a factor is acting simultaneously over a wide range of special habitats within a major habitat or community type, for example the effect of aerial spraying on the fauna of a mixed woodland. If a fixed number of samples of the fauna are taken - say from three different zones of a stream: riffles, pools and marginal areas -it will be found that the first few samples will provide a more complete picture of the total observed fauna in one zone than in the other. In other words because of the different diversities and microdistributions of the fauna in the different zones, and also possibly because of the differing efficiencies of the collecting methods under these conditions, the rates of accumulation of new species vary from zone to zone. A zone in which a high percentage of the total species taken in a large number of samples *(n)* were recorded in the first few samples, can have a smaller number of samples drawn from it than a zone in which the rate of accumulation is less rapid.

A method of calculating the average number of new species contributed by the Kth successive sample (when Kcan equal any number from 1 upwards) has been devised by Gaufin, Harris & Walter, who also provide a table of the coefficients necessary for up to a total of ten samples. Their formula is:

$$S_K = \frac{1}{S} \sum_{i=1}^{n-K+1} a_i KS_i$$

where S_K = the number of new species added by the Kth sample, n = the total number of samples taken in the preliminary survey, S = the total number of species taken in the n samples, S_i =the number of different species appearing in i out of n samples and a, K are coefficients by which successive values of S; are multiplied in calculating the summation item. By summing a series of values of S_K (for $K = 1$ to a predetermined number K') it is possible to compute the average number of species found in K' samples. A method of calculating the standard errors of such estimates has been developed by Harris.

BIOLOGICAL AND OTHER QUALITATIVE ASPECTS OF SAMPLING

Hitherto stress has been laid on the statistical aspects of sampling; however, there are certain biological problems that are of cardinal importance. The ecologist should always remember that the computation of fiducial limits of estimates only tells him the consistency of the samples he has collected, and the statistical techniques cannot be blamed if these have been consistently excluding a major part of the population. This is indeed the most serious biological problem in all sampling work, namely to ensure that there is not a part of the population with a behavioural pattern or habitat preference such that it is never sampled.

Less dangerous because the phenomenon will be recorded, although it may well be misinterpreted, is the tendency for the behaviour of the animal to change and affect its sampling properties. The reactions of the fly *Meromyza* to weather conditions and of the larvae of different species of corixid water-bug to light lead to errors when net sampling is carried out under differentconditions. In many, perhaps almost all, insects the behaviour alters with age, and this may lead to a confusion between a change in behaviour and death. For example older females of the mirid grass *bug, Leptopterna dolobrata,* spend more time on the base of the grass shoots and less on the tops than do males. If a marking and recapture experiment is carried out and the bugs collected by sweeping, the females will appear to be much shorter-lived than the males. But this is because as they age, the females enter the sampling zone less frequently; they are actually longer-lived than the males.

The variations in the distribution of the insect may well be linked with some character of the environment or its host plant, and it is important to distinguish this variation from sampling variance otherwise the fiducial limits of the estimates may become so wide that no conclusions can be drawn, that is systematic errors will arise. When population trend is eliminated, approximately 50 % of the variance of the counts of populations of the sugar cane froghopper *(Aeneolamia varia),* on sugar cane stools could be accounted for in terms of the size of the stools; hence more 'accurate' estimates would be obtained by taking stool size into consideration and adjusting population estimates, or by allowing for it in the analysis. This is particularly important when comparisons of froghopper numbers are being made between insecticide-treated fields;

the fields are of different ages and hence with stools of different sizes. Many other examples could be given of environmental factors influencing distribution; they are commonly microclimatic in nature, e.g. soil moisture affects the distribution of the cocoons of a sawfly. Even when the environment appears uniform distribution may be systematically non-random; for example, the most dense populations of the beetle, *Tribolium,* in flour are always adjacent to the walls of the container. Covariance analyses are often helpful in such situations.

In most habitats a random sample can be selected by numbering the habitat on a grid system and using a table of random numbers; an acceptable approximation is to move a certain number of units determined by a random number along one side of a plot and then turn at right angles and move a second number of units into the plot. Such a method can be used for soil samples, crops and herbage, aquatic samples and, with some modifications, with trees.

The studies of Howe on stored grain insects illustrate some of the problems of random sampling. The bag of grain may be selected randomly or the suction probe may be inserted at a random point (the depth of sampling is usually an important biological component of the variability and hence each depth band tends to be sampled separately). The quantity of grain obtained in this way is often more than can conveniently be examined for insects, and thus a subsample must be taken. Howe has compared four ways of subsampling grain and he has shown that two mechanical separators, a machine devised for the separation of fine granular fuels and the standard apparatus for splitting grain samples for grading, and one hand method gave reliable randomized subsamples. The hand method consisted of pouring the grain over the spherical bottom of a short-necked flask upturned in a glass dish; the grain becomes more or less evenly distributed round the dish and samples of a fixed volume are scooped up radially, the dish being turned through a random angle between each scoop. Although this method was found reliable by Howe, it does depend, as he stresses, on the personal skill and avoidance of bias by the operator.

10

Absolute Population Estimates Using Marking Techniques

Studying plaice and waterfowl populations respectively, Petersen and Lincoln independently developed a marking method from which the total population may be estimated. This was based on the principle that if a proportion of the population was marked in some way, returned to the original population and then, after complete mixing, a second sample was taken, the number of marked individuals in the second sample would have the same ratio to the total numbers in the second sample as the total of marked individuals originally released would have to the total population. As the first three quantities were known the latter could easily be calculated. This method has been extensively developed and provides the major alternative absolute method to those based on the count of animals within a fixed unit of the habitat; it has the advantage that its accuracy does not depend on an assessment of the number of sampling units in the habitat and, as has already been stressed, it is a wise practice to use more than one method simultaneously. There are also certain other methods of estimating a population that depend on the presence of marked individuals, but use a different principle to the Lincoln Index. A comprehensive survey of these methods is provided by Seber.

A basic prerequisite to the use of these methods of population estimation is a technique for marking the animals so that they can be released unharmed and unaffected into the wild and recognized again on recapture. Such techniques may also be used in studies on behaviour, e.g. dispersal, longevity and growth; but for convenience all aspects of marking insects and other invertebrates will be discussed here.

METHODS OF MARKING ANIMALS

A fundamental requirement of any marking technique is that it shall not affect the longevity or behaviour of the animals. An attempt should always be made to confirm that this is true in the particular case under investigation because, for example, although the pigments used in most markers may be non-toxic, the solvents are often toxic. This may be checked in the laboratory or field cage by keeping samples of living marked and unmarked individuals and comparing longevity or in the field by comparing the longevity of individuals bearing differing numbers of marks. Newly emerged insects may be more sensitive to the toxic substances used in markers than older insects and the attachment of labels to their wings, which has no effect on old insects, may cause distortion due to interference with the blood circulation.

It should also be borne in mind that the presence of conspicuous marks may well destroy an animal's natural camouflage and make it more liable or, as Hartley observed with marked snails, less liable to predation. This effect is difficult to assess; it can to some extent be avoided by marking in inconspicuous places or by the use of fluorescent powders, dyes in powder form, phenolphthalein solution or radioactive tracers whose presence is only detectable by the use of a special technique after recapture. The effect on predation of a conspicuous, but convenient, marking method could be measured by marking further individuals with one of these invisible methods and comparing longevity. The effect of a conspicuous mark can be checked through choice experiments in the laboratory: Buckner confirmed that small mammals are not influenced if their prey was stained with vital dyes.

Another aspect of the conspicuous mark is that, where the animals are sampled by a method that relies on the sight of the collector, then the marked individuals may tend to be collected more than unmarked ones.

A third problem concerns the durability of the mark; some

paints, particularly cellulose lacquers, may flake off leaving the animal apparently unmarked; student's oil paints and powdered dyes may wash off; some fluorescent powders may lose this property on exposure to sunlight or be abraded during collection and a radioactive isotope could decay and/or be excreted by the animal. Immature animals will lose marks on their cuticle when they moult. Laboratory tests of durability are not always reliable; Blinn found that cellulose lacquers would remain on the shells of land snails for two years in the laboratory, but they only lasted about one in the field, and the rate at which a radioactive isotope is lost may depend upon the animal's diet and other factors.

The handling and release of marked individuals may also affect their subsequent life expectancy and behaviour; these problems are discussed below.

The amount of effort that can be put into a marking programme (i.e. its cost) will be related to the percentage of recoveries that can be expected; a high cost per individual marked will be justified where the recovery rate is high.

Group Marking Methods

These methods enable a large number of animals to be marked in the same

way and are perfectly adequate for most capture-recapture population estimations and in dispersal studies. Almost all the methods are capable of one or two variants so that two or three groups may be marked differently, and thus the distinction made by Dobson (1962) between group methods and common marking methods in which all individuals are marked in the same way is not followed. Marking methods have also been reviewed by Dobson and by Gangwere, Chavin & Evans (1964), who give further references.

Paints and Solution of Dyes

Materials

Artist's oil paint is perhaps the most extensively used marking material; it can, of course, be obtained in a variety of colours and has been used successfully for marking moths, tsetse flies, bed bugs , locusts and grasshoppers; mirids, flies, beetles and others. However, Davey found them toxic to certain locusts, although this may have been the effect of the dilutant and they are, of course, slow drying. Artist's poster paints have been used to mark mosquitoes .

Nitrocellulose lacquers or paints (e.g model aircraft dope) and alkyl vinyl resin paints are quick drying and have been used by a number of workers; on snails, where they were applied to a small area on the underside of the shell, from which the periostracum had been scraped, and on grass hoppers, although the first-named authors found them less satisfactory than artist's oil paints. They have also been applied to ants, lace bugs, dragonflies, mites, tipulid flies and carabid beetles and mosquitoes. Fluorescent lacquer enamels (e.g. 'Glo-craft' or 'Dayglo') or fluorescent pigments with gum arabic glue plus a trace of detergent have been used to mark tsetse flies, chafer beetles and lepidopterous caterpillars. Animals marked in these ways may be spotted after dark in the field at distances of up to 25-30 ft (8-10 m) by the use of a beam of ultraviolet light; such a beam may be produced by a battery-powered lamp.

Reflecting paints may also be used to mark animals for detection at night and have the advantage that they can be seen with a small (12-V) hand torch for up to 30 ft (10 m); Rennison, Lumsden & Webb used this method with tsetse flies and found that a mixture of the minute glass beads in a thin varnish of shellac or in a gum solution with a trace of detergent was preferable to the commercial aerosol form of the paint (e.g. Codit, 7211, reflecting paint). Aluminium paint was found to adhere well to scraped areas of the elytra of carabid beetles.

Aniline dyes dissolved in a mixture of alcohol and shellac or in alcohol alone have been used, respectively, to mark cucumber beetles, *Diabrotica* and gypsy moths, *Lymantria*. Solutions of various stains in alcohol, such as eosin, orange G and Congo red, have been used to mark adult Lepidoptera; petroleum based inks (e.g. Easterbrook Flowmaster) have been found particularly useful, staining the integument below the scales. Fluorescent dye solutions, in alcohol or acetone, principally rhodamine B, have been used to mark mosquitoes and *Drosophila* and the tick, *Argus*. Working on house flies Peffly & Labrecque used a 6 % solution of phenolphthalein in acetone; the marked flies were identified on recapture by placing them in 1 / sodium hydroxide solution, whereupon they became purple. Fales *et al.* used waterproof inks to mark face flies.

Application

When the paints or solutions are in their most concentrated form they are most conveniently applied by the use of an entomological pin, a sharpened match-stick, a single bristle, or even a fine dry grass stem. With quick-drying cellulose lacquer it may be necessary to dilute them slightly with acetone or another solvent; if this is not done a fine skin may form over the droplet on the pin and it will not adhere firmly to the animal. With artist's oil paints, dyes in solution or diluted lacquers a camel-hair brush may be used, although generally this method has no advantage over the use of a pin and frequently leads to the application of too large a mark. If the mark covers any of the sense organs or joints the specimen will have to be discarded. Freeman found a fine syringe was suitable for applying cellulose lacquers to tipulid flies. Felt tip pens provide a convenient way for marking some large insects.

The paints or solutions may be further diluted with acetone, dilute alcohol or other solvents and sprayed on. This may be done with a hand atomizer (e.g. a nasal spray) while the insects are contained in a small wire cage; mortality during marking by this method can be reduced if, immediately after spraying, the insects are quickly dried in the draught from an electric fan. Large numbers of moths may be rapidly marked, the number marked being recorded by a cyclone transfer machine and associated photoelectric cell described by Wolf & Stimman. This technique has been extended to field marking of locusts and butterflies by the use of a spray gun and an oil can (e.g. 'Plews Oiler') respectively; it was found that individuals could be marked at a distance of 15 feet (5 m) or more. Davey showed that for the same cost (labour and time) nearly ten times as many locusts could be marked by this method than by that involving the capture and handling of each individual, and Davey & O'Rourke found that handling itself could have fatal effects on tabanid flies, *Chrysops*.

Dyes and Fluorescent Substances in Powder Form

Hairy insects may be marked by dusting them with various dyes in powder form; this is most easily done by applying the dusts from a powder dispenser (insufflator) or by producing a dust storm in a cage with a jet of air [e.g. produced by a bicycle pump. Apparatus that could be used for this purpose is described by Dunn & Mechalas and Frankie. Only a very small quantity of powder is necessary.

Non-fluorescent dyes that have been found useful are the rotor and waxoline group. The marked insects are recognized by laying them on a piece of white filter paper and dropping acetone on to them, when a coloured spot or ring forms beneath those that have been marked. As the testing involves the killing of insects this method is not suitable for extensive recapture work, and laboratory tests showed that with blowflies the mark may only last for one week and seldom for more than two. However, dyes of two different colours may be applied to the same insect and distinguished in the spotting. This method has been used for calypterate flies by Schoof.& Mail, Quarterman, Mathis & Kilpatrick and MacLeod & Donnelly and for the frit fly *(Oscinella frit)* (Southwood & Jepson, unpublished) and various mirids. It is possible that under certain circumstances it could be used for marking large aggregated populations in the field.

Fluorescent substances (e.g. zinc sulphide powders, especially 'Helecon' nos. 1757, 1953, 2200, 2225, 2267 & 3206), whose presence is detected by placing the animals under an UV lamp, have also been used extensively for marking. Multiple marking is possible, although combinations sometimes produce distinct fluorescences, and because they may be detected without killing the insect, they are suitable for use in capture-recapture population estimation. Although with the stable fly, *Stomoxys,* field cage experiments suggested that they could adversely affect longevity, this was not detected in trails with *Drosophila*. Lists of the various materials that may be used are given by Staniland and Bailey, Eliason & This. They may be applied directly, but better adhesion can be obtained by mixing one part of the dye with six parts of gum arabic, adding water until a paste is formed, then drying the paste and pulverizing it in a mortar. This powder is applied to the insects which are then placed in a high humidity; the gum arabic particles absorb sufficient moisture to make them adhere to the insect. This method has been used for marking mosquitoes by Reeves, Brookman & Hammon. The movements of foraging bees have been studied by marking them with a fluorescent powder as they leave the hive; this is conveniently done by forcing them to walk between two strips of velveteen liberally dusted with the marker; tabanids caught in a canopy trap have been self-marked a similar way. Bees have also been marked when visiting flowers by dusting these with a mixture of the fluorescent powder and a carrier such as talc or lycopodium dust. The bees leave a trail of powder, which can be detected after dark with an UV lamp, on the other flowers they have visited. Fluorescein and rhodamine B have been founded useful in this work; all the bees leaving the hive are marked and the marks last for weeks. The possibility that unmarked individuals may bear a few particles that will fluoresce under UV light should be remembered when using these markers. Wild caught mosquitoes have been found with fluorescent blue, purple, green, white, yellow and orange spots, therefore the use of rhodamine B, which fluoresces red, was recommended.

An ingenious self-marking method for newly emerged calypterate flies has been devised by Norris. The principle is that the soil or medium which contains the fly puparia is covered with a mixture of about forty parts to one of fine sand and fluorescent powder or the puparia are coated with dye; as the flies emerge a small quantity of the dust adheres to the ptilinum; and when after emergence the ptilinum is retracted the dust becomes lodged in the ptilinal suture. Sometimes in an examination of the faces of such marked flies in UV light this suture will be seen to shine vividly; but a more reliable technique is to crush the whole head on a filter paper, at the same time adding a small amount of the appropriate solvent: the mark may be seen on the paper, using UV if necessary. Many types of fluorescent powder have been found satisfactory: 'Lumogen', 'Tinopal' and 'Day-glo'. Oil soluble dyes, e.g. Calco red, have also been used with fruit flies (Steiner).

Labels

Bands and rings are used extensively in work on birds and mammals, but the small size of most insects usually precludes these convenient methods. Butterflies and locusts have, however, been marked by attaching small labels with a word or a code written in waterproof black ink to part of their wings; in the Lepidoptera the area should first be denuded of scales. Earlier workers used paper or cellophane stuck on with an adhesive, such as 'Durofix', but recently 'Sellotape' has been found satisfactory. Klock, Pimentel & Stenburg devised a machine that glued lengths of coloured thread to anaesthetized flies. Punched ferrous labels have been used on bees; they may be magnetically removed. It is possible that large-bodied insects might be tagged internally using the methods of fishery workers. Great progress has also been made with vertebrates through the use of radio-telemetry; developments in electronics may make it possible for the largest insects to carry a transmitter so this technique could be used.

Mutilation

This method is also more widely used with vertebrates, especially fish, amphibians and reptiles, than with the smaller insects, for a mark in order to be easily visible may be proportionally so large as to affect the insect's behaviour. Lepidoptera have been marked by clipping their wings, carabid beetles by damaging their elytra in various ways: incising the edges, punching or burning small holes or by scraping away the surface of the elytra between certain striae; the latter method can be used with large chrysomelids (Southwood, unpublished). Crabs have been marked by cutting some of the teeth on the carapace and orthopteroids by notching the pronotum and amputating tegmina.

Marking Internally by Injection

If the tissues of the animal can be marked in some way this

has the great advantage with an arthropod that the mark is not lost during moulting. It has been found possible to mark crayfish by injecting a small amount of 'Bates numbering machine ink' into the venter of the abdomen; the black, blue and red inks were found to be non-poisonous; similarly indian ink may be used in the fish, *Gambusia*. This method might be applied to other large arthropods that have an area of almost transparent cuticle. Attempts have been made to mark adult mosquitoes by feeding the larvae with dyes; these have mostly been unsatisfactory, often leading to high mortalities.

Marking by Feeding with Dyes

The marking of invertebrates with a vital dye incorporated in the food, is a valuable tool: its significance is enhanced if the mark is retained from larval to adult life (see section 8 below) and if it can be detected without killing the animal. Many workers have screened a wide range of dyes, but few have been found satisfactory: the majority are either rapidly excreted or prove toxic. Calco oil red if fed to immature stages, has been found to mark adults and sometimes the resultant eggs (but not the F_1 larvae) of several cotton insects *(Heliothis, Platyedra, Anthonomus)*. The dye is fed in a natural oil (e.g. cotton seed) with the larval diet at a concentration of about 0.01 % dye/unit diet: it is detected in the adult by crushing the abdomen. Similar results have been obtained with oil soluble 'Deep Black BB' and 'Blue II'.

Sawfly larvae have been marked by feeding them, in the last instar, on foliage treated with solutions of rhodamine B (series 4; 3.7 g/1) and Nile Blue Sulphate (series 5; 0.4g/1). The cocoons, adults and eggs were all marked and the dyes were visible externally; rhodamine B fluorescing bright yellow -orange under UV. Mosquitoes and the eyes of a small fish have been marked by feeding them on a 0.01 % solution of rhodamine B in a sugar solution ; the same dye has also been used to stain the gut of *Drosophila* . House flies have been marked with thiorescin and fluorescein; in some cases only a small proportion of the insects could be induced to feed on the solution. South marked slugs by feeding them on agar jelly containing 0.2 % neutral red; the digestive gland became deeply stained and the colour was easily visible through the foot. Bloodsucking Diptera have been marked by allowing them to feed on a cow which had had 200 ml of an aqueous solution, containing 4 g of trypan blue, intravenously administered over 20 minutes. The dyestuff can be detected by a paper chromatographic technique, in which the gut contents of the fly is mixed with 0.1 N sodium hydroxide solution and applied to a narrow strip of Whatman No. -1 chromatographic filter paper, which is developed in 0.1 N sodium hydroxide solution; the trypan blue remains at the origin, other marks due to the gut contents move away. Haematophagous animals may also be marked with specific agglutinins. These dyes or stains may often be detected in the faeces of predators that have fed on marked animals.

Genes, Mutant and Normal

Dispersal may be studied by the use of mutant genes or when various genotypes are clearly distinct, by their different proportions in adjacent colonies; however, it should be remembered that selection may operate differentially in the different colonies, perhaps on young stages before the genotype becomes identifiable. Different sexes and age classes also provide naturally marked groups.

Rare Elements

Many elements, when exposed to a source of neutrons, become radioactive and emit a characteristic spectrum of gamma-rays. The process is known as neutron activation. Various workers have marked animals by incorporating rare elements in them and subsequently recognizing the mark by neutron activation and gamma spectroscopy. The great advantage of the method is that, theoretically at least, the mark may be retained from larval to adult life and self-marking is possible. For example if a small quantity of a rare element was mixed into a mosquito's breeding pool any adults emerging should be detectable any time in their life: the contributions of different breeding sites to a population could be determined. The disadvantages are that the equipment is extremely expensive (but is often available in nuclear physics or engineering centres); the procedure for gamma spectroscopy for rare elements is fatal for the animals and there are often difficulties, . undoubtedly of a basic physiological nature, in getting an adequate quantity of the rare element absorbed into the body (rather than just held in the gut) and retained in the tissues.

Dysprosium, first used by Riebartsch has been found to be a life-long marker for some Lepidoptera and *Drosophila* ; rubidium for the cabbage looper, *Trichoplusia ni* larva, europium for Lepidoptera and manganese for fruit fly, *Ceratitis*. Rubidium when sprayed on host plants is readily absorbed; Berry *et al.* found a straight line relationship between the concentrations of rubidium in foliar spray and in the male moths that developed from the larvae that had been reared on the sprayed plants; but with the pea aphid, *Acyrthosiphon pisum,* the biological half-life was only about one day and detectable quantities only remained for four days after the aphid had left the plant. Ito showed that incorporation of europium (44 or 80 ppm) into the diets of moth larvae, *Hyphantria* and *Spodoptera,* did not affect growth and survival.

Alternatively the rare elements may be used as labels. Cerium is absorbed into the cuticle of insects; Rahalkar *et al.* used it (50 ug $CeCl_3$/insect in an alcoholic solution) to mark weevils, *Rhynchophorus,* by topical application on the elytra, and Bate *et al.* used a similar method with gold for the elm bark beetle. Jahn *et al.* obtained marking by spraying dysprosium and europium salts, with a little detergent, on to the insects.

Emerging calypterate diptera may self-mark by contamination of the ptilinal suture. Various dyes are often used for this purpose , rare elements may be even more effective. Haisch, Stark and Forster marked emerging *Rhagoletis* with dysprosium and samarium, mixed with fine sand and silica gel (1 % by wt.) at concentrations of 0.1 % by weight. Provided the pupa were 3 cm deep the error in self marking was about 2 %.

Radioactive Isotopes

It is impossible to provide a full bibliography for this extensively used method; one is however given by Anon. and many papers are listed in reviews by Jenkins & Hassett, Lindquist, Hinton, Pendleton.

TABLE 10.1 : SOME CHARACTERS OF THE PRINCIPAL ISOTOPES USED IN ECOLOGICAL RESEARCH (*USED AS LABELS)

Isotope	Symbol	Half-life (*approximate*)	*Radiations and energies (MeV)* Beta (*maximum*)	Gamma
^{14}Carbon	C	5760 years	0.16	
^{137}Cesium	Cs	30 years	0.31	0.66
* ^{60}Cobalt	Co	5.27 years	0.31	1.2
* ^{65}Zinc	Zn	245 days	0.33	1.1
*^{195}Gold	Au	185 days	0.1	
^{45}Calcium	Ca	165 days	0.25	
* ^{182}Tantalum	Ta	115 days	0.51	range, max. 1.2
^{35}Sulphur	S	87 days	0.17	
*^{46}Scandium	Sc	84 days	0.36	0.89
* ^{192}Iridium	Ir	74 days	0.67	0.6
^{89}Strontium	Sr	55 days	1.50	
^{59}Iron	Fe	45 days	0.46	1.2
^{32}Phosphorus	P	14.2 days	1.71	
^{131}Iodine	I	8 days	0.61	0.36

The radioactive isotopes of elements are unstable and they disintegrate emitting radiations and forming other, usually non-radioactive, isotopes; the rate of disintegration is characteristic for each isotope and is described in terms of its half-life. Of the three types of radiation, entitled alpha, beta and gamma, the isotopes used in ecological work produce only the two last named. Beta particles have less power of penetration than gamma rays; however, the actual energies (expressed as MeV = million electron volts) of the radiations differ for the different isotopes. Radioactivity is measured in units of curies (1 curie = 3.7×10^{10} disintegrations/second) and the actual mass of chemical element constituting one curie will vary depending on the half-life of the isotope. The specific activity of any material or solution gives the relationship of the amount of radio-isotope to the total element content and is expressed in terms such as c/g. In biological work smaller units are usually required and these are: millicurie (mc) = 10^{-3}c and microcurie (pc) = 10^{-6}c. Further basic information on isotopes may be found in textbooks, e.g. Comar, Francis, Mulligan & Wormall , Faires & Parks, O'Brien & Wolfe Thornburn, or in tables, e.g. Hollander, Perlman & Seaborg, Radio Chemical Manual. Table 10.1 gives the characters of the principal isotopes used in entomological work. The use of isotopes in ecology is by no means confined to marking; they may also be used in studies on predation and energy flow. Great care should be exercised in all work with radioactive isotopes and precautions taken not to contaminate the environment (and the operators!). For this reason long half-life isotopes should normally be avoided in fieldwork.

There are basically two methods of marking animals with isotopes, although the dividing line is not sharp. The isotopes may be used as a label outside the animal or alternatively they may be fed to the animal and incorporated in its tissues. Which method and which isotope will depend on the precise nature of the work. If it is desired to trace the animal's movements, in the soil or in wood, or to locate it from a distance, then it is the penetrating gamma radiations that will have to be detected.

If the object is to mark part of the population in a way that can be recognized after the animal has been recaptured, then a wider range of isotopes are available, including those of carbon, calcium, phosphorus and sulphur, which, if fed to the animal, are readily incorporated into its tissues. This marking method is preferable to the use of radioactive labels and, indeed, to many other marking methods in that animals can be tagged with the minimum amount of manipulation and the mark is invisible to predators. It is also almost unique (see above) as the mark will not be shed when the insect moults; this character gives the method great potential for the estimation of population size and mortality in immature stages and may be passed on to the next generation. It is necessary, however, to choose an isotope (e.g. sulphur) with a reasonably long half-life (unless the insect's life-cycle is very short) and to carry out laboratory tests on the extent to which the radioactive isotope atoms are excreted by the animal and replaced by normal ones.

Ants pass food to other members of their colonies by trophallaxis, whilst some gregarious insects (e.g. *Triatoma)* feel intra-integumentally on the gut contents of others. With these the risk of secondary tagging becomes significant, thereby invalidating the use of marking for Lincoln-Petersen Index type estimates; they may however be used, with ants, to determine information on foraging area. However, Stradling has shown that if the ants are starved for four days after marking, before return to the nest,32P becomes incorporated into the tissues and secondary tagging does not occur.

Labels

Besides acting as a suitable gamma source, radioactive labels need to be nontoxic and durable. They may be attached externally, when corrosion or loss are major problems, or inserted into the body, when toxicity or disturbed behaviour present difficulties. Among the elements used are cobalt (^{60}Co), tantalum ($^{1\ 82}$Ta), iron (^{59}Fe) and iridium (^{192}1r). For protection, iron may be used in stainless steel wire, iridium in a platinum-iridium wire and cobalt can be gold plated. A benzene soluble DeKhontinsky type cement has been found a good glue for attaching labels. Alternatively, cobalt in the form of the nitrate can be made up with a water-resistant resin glue (e.g. 'Bond Fast') at a strength of abour 1.6 me/ml and applied to the surface of the insect; likewise ^{192}1r, as the chloride, mixed with some cellulose paint and rubber solution or ^{6}5Zn, as the oxide, mixed with cellulose paint and applied with a special applicator. Minute labels (about 0.05 mm × 0.16 × 0.23 – 0.46 mm) of metallic tantalum have been used to mark coccinellid larvae and pentatomoid bugs, being attached by seccotine glue or cellulose paint, and to mark earthworms, introduced into the coelomic cavity. The initial specific acitivity of the tantalum labels used by Banks *et al.* on the bugs was about 8c; with the much smaller young coccinellid larva, labels with an average activity of 7.5 μc produced marked adverse effects and those used in the field work had a lower specific activity (1.3 μc). Workers using cobalt have applied labels with much higher specific activities, and Sullivan has shown that if this is greater than 50 μc the survival of the weevil, *Pissodes strobi,* is affected; this does not, however, become apparent for several months. But I am doubtful of the validity of the implied claim of some workers that, if the dosage is just small enough for the mortality to be negligible, then the animals' behaviour is normal. The problem of tagging and maximum total radiation dose is discussed fully by Griffin; for the ecologist it is a good principle to use the minimum radiation dose consistent with detection, as was done by van der Meijden, who showed that the pupation success of moth larvae was not affected by labels containing 3-4 μc of ^{192}Ir: these labels could be detected from about 60 cm.

Submerging the animals in a solution containing the radioactive isotope is another method that may be classified as labelling, although it is possible that a small amount of the isotope could become incorporated in the tissues. This method was first developed by Roth & Hoffman, who marked house flies, wasps, leafhoppers, grasshoppers and coleoptera by dipping them for one minute into a solution containing ^{32}P at the concentration of 5 μc/ml together with a wetting agent. The marker did not penetrate the body of the animal, yet it could be removed only by prolonged washing. Large numbers of insects may be marked quickly by this technique and it has frequently been used; some examples are:

Animal	Isotope	Strength of solution applied	Authors
Boll weevi, *Anthonomus*	^{60}Co	0.5 μc/ml	Barbers *et al.*, 1954
Bug, *Eurygaster*	^{60}Co	12 me/ml	Rakitin, 1963
Pine weevil, *Pissodes*	^{46}SC	*c.* 7.5 me/ml	Godwin *et al.*, 1957
Ant, Lasius	^{32}p	*c.* 1 μc/ml	Odum & Pontin, 1961
Tick, Amblyomma	^{32}P	10 μc/ml	Knapp *et.*, 1956
Tick, *Amblyomma*	^{59}Fe	10 μc/ml	Smittle *et al.*, 1967

The strengths of the solution applied varied greatly; it is possible that in some instances activity levels well above the minimum necessary were used. Roth & Hoffman found that the isotopes on their insects could be detected for about eighteen days and Godwin *et al.* could recognize their marked weevils for five months; the difference largely stemming from the half-lives of the isotopes. The marking can. conveniently be done in a covered funnel, with a constricted neck or straining plate to prevent the insects being washed out. More delicate animals cannot be marked by this method and because of the 'violence' involved it is doubtful if it should be used even with robust species where an alternative technique is available.

A few instances are recorded of marking insects by spraying them with radioactive isotopes; this is, however, extremely hazardous for the operator and cannot be recommended.

Individual Marking Methods

If each individual can be separately marked additional information can be obtained on longevity and dispersal and, if they

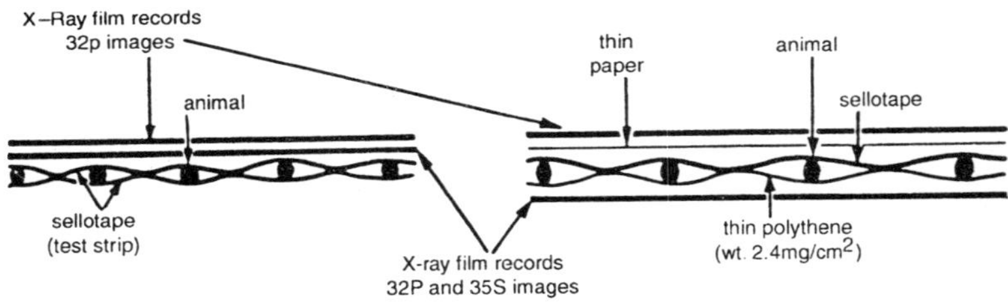

Figure 10.1 : ***Double exposure autoradiographic methods of discriminating between animals marked with 32P and 35S-diagrammatic sections of arrangement of test strip and films in cassette:*** **a.** ***method of Gillie;*** **b.** ***method of Lewis & Waloff.***

can be aged and sexed initially, survival can be related to these and other characters. Birth- and death-rates may be more easily calculated and the excessive handling of animals, recaptured more than once, with its attendant problems, may be avoided. It may also, for example, be possible by weighing individual females to assess the rate of oviposition in the field, and it provides a method for assessing the randomness of recapture. The extent to which

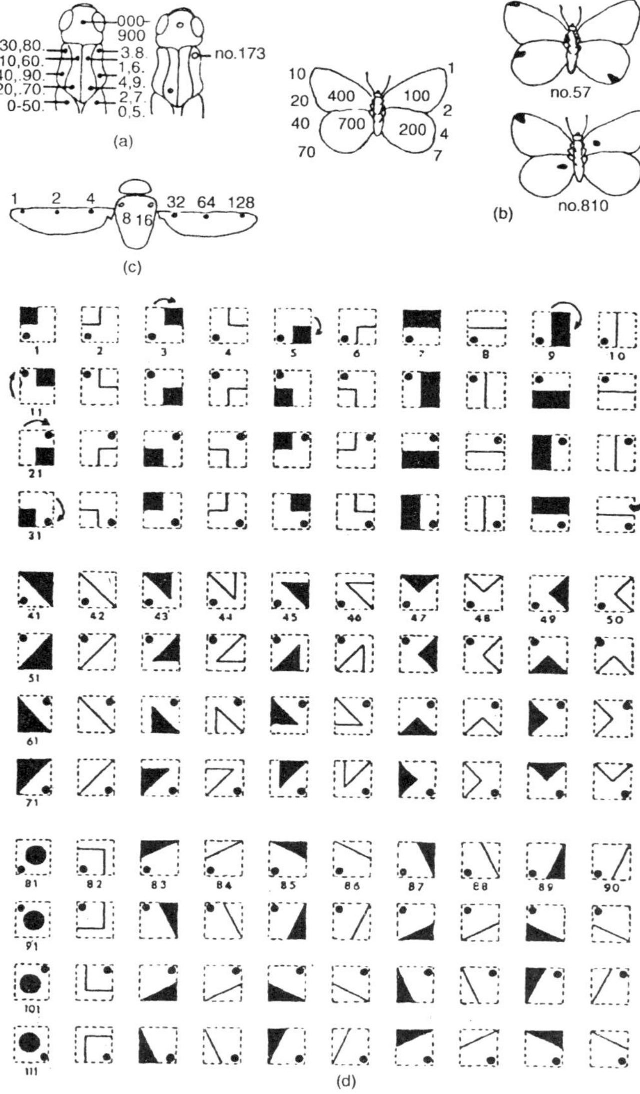

Figure 10.2 : ***Systems for marking insects individually using colour and postion codes; a. Richards and Waloff's 'decimal system'; b. Brussard's modification of Ehrlich and Davidson's '1—2—4—7 system'; c. Sheppard et al.'s 'binomial system'; d. Whites's 'shape code'.***

the higher cost, in terms of effort, of marking animals individually is justified will depend on the percentage of the marked individuals that are recovered; high recovery rates justify elaborate marking programmes; with low recovery rates individual marking is seldom justified.

If small labels can be attached to the insects' wings (see above) there is no problem in marking each individual, and Nielsen records that in the laboratory he marked butterflies individually on the wing with a rubber stamp. With most small invertebrates individuals marks have to be obtained by a combination of spots in various positions, the numerical range of the coding often being increased by the use of various colours. It is generally wise to follow the policy of Michener *et al.* and ensure that all individuals bear the same number of marks, so that if one is lost number 121 doesn't become say 21, but is immediately recognised as an 'unreadable' mark. It is also desirable to use the minimum number of colours in marking any individual insect, as the change in colours increases the handling time. Obviously a change in colour between say the ranges 1-99 and 100-199 is less of a practical problem, than the use of two colours in a single mark. The actual pattern will depend on the size and shape of the insect, the number of colours available and the number of individual marks required. Patterns for bodies and wings have been devised for dragonflies, bed bugs, tsetse flies, bees, grasshoppers, snails, craneflies, mosquitoes and lepidoptera

The system of Richards & Waloff is logical and versatile, enabling up to 999 individuals to be marked with continuous numbering. On the right of the thorax is a spot that represents the unit, on the left one that represents the tens, the head spot the hundreds; 1-5 and 10-50 are white spots, 6-0 and 60-00 are red; ten different colours are used for the head mark including the zero class (for 199). With this system the addition of a single further mark would allow another 10,000 individuals to be numbered. With smaller insects the number of spots can be reduced by increasing the number of colours used, as Richards & Waloff did with the hundred mark; however, the practical problem of switching rapidly between more than three colours is serious. The rapid reading of the mark is another advantage and Richards & Waloff's code is particularly clear in this respect because the same colours and corresponding positions represent the same figures in units and tens.

Richards & Waloff's code may be considered a decimal code, based on tens. Ehrlich & Davidson developed a 1-2-4-7 marking system, modified by Brussard to mark up to 1000 individuals with one colour. A binomial system for marking mosquitoes was developed by Sheppard *et al.*, up to 255 individuals can be marked with a single colour but as with Brussard's method the number of marks is variable and the quick and accurate reading of the mark a matter of experience.

A more elaborate code of shapes has been devised by White. The system is self-checking (if marks are lost), a single colour may be used up to 160, but as White points out rapid reading becomes more difficult beyond 120.,

If the body and/or wings are unsuitable for spotting the legs may be marked, but the removal of such marks by the animals during cleaning movements is a real risk. Kuenzler was able to number lycosid spiders by marking their legs with white enamel paint and Corbet hippoboscid flies by marking both the body and the legs.

Mutlilation may be used as an alternative to colour marks. By allocating a code of numbers to the teeth on a crab's carapace it is possible to number up to about forty individuals (depending, of course, on the age and species of crab) by cutting off one or more teeth. In the same way by numbering the interstrial spaces of a carabid, a patch in any one of which could be scraped, approximately 999 individuals could be numbered. Gangwere *et al.* developed a method of individually marking orthopteroids by notching the pronotal margin and amputating the tegmina.

Handling Techniques

If the animals are to be marked by spraying or dusting this can be done whilst they are still active, either in the field or in small cages (see above); but for marking with paints, especially precise spotting, and for most methods of labelling and mutilation, it is necessary for the animal to be still. Often it can be held between the fingers; where this is impossible the animal must be held by some other methods, anaesthetized or chilled.

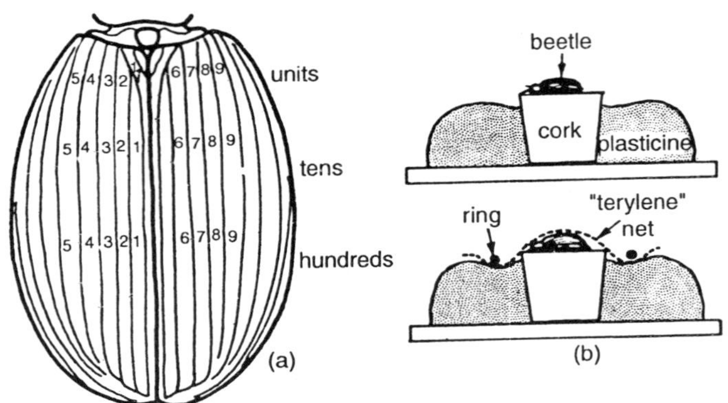

Figure 10.3 : ***a, A 'mutilation system' for marking carabid beetles individually by scraping certain positions on the elytra, applied to*** **Agonum fuliginosum. b.** ***A 'device for holding a hard-bodied animal whilst it is marked.***

The animals may be held still under a net or with a hair. One end of the hair is fixed (e.g. in seccotine) and the other has a piece of Plasticine on it; the animal is placed underneath and the hair may be tightened under a stereoscopic microscope by pressure on the Plasticine. More robust insects may be held immobile on top of a cork by a piece of Terylene net which is kept taut by a ring (of about 2 in. diameter, but of course depending on the size of the animal); the ring is pushed down into the surrounding Plasticine and the animal can be marked through the holes of the net. Conway, Tripis & McClelland held mosquitoes for spot marking between diaphragms of nylon mesh and stockinet. These devices may be placed under a microscope.

Insects can also be held by suction; Muir found that a mirid bug was conveniently held by the weak suction of a pipette, formed from the ground-down point of a coarse hypodermic needle, connected to a water pump. The picking up of the bugs by this weak suction was facilitated by coating the inside of the glass dish, in which they were held, by an unsintered dispersion of 'Fluon' GPI, polytetrafluoroethylene (PTFE), which presented an almost frictionless surface. Butterflies may be held in a clamp.

If none of these methods can be used or if the insects are so active that they cannot even be handled and counted they may have to be immobilized. Chilling at a temperature between 1 and 5°C is probably the best method and may be done in a lagged tank surrounded by an ice-water mixture. Alternatively an anaesthetic may be employed: carbon dioxide is often used and is easily produced from 'dry ice'. The last mentioned author gives the following table of recovery times of house flies from carbon dioxide anaesthesia at 21° C (70° F):

Exposure time	*Recovery time*
Up to 5 min	1-2 min
5-30 min	3-5 min
30-60 min	5-10 min

Working on the earwig, *Forficula,* Lamb & Wellington found that although activity may be quickly resumed after carbon dioxide anaesthesia of less than 1 minute, normal behaviour. could be affected for many hours; they recommend a 24-hour recovery period. Dalmat found that female blackflies (Simulidae) laid more eggs than normal after being subjected to carbon dioxide, but Edgar found this did not affect the development of the spider, *Lycosa lugubris.* Other anaesthetics are ether, chloroform, nitrogen and nitrous oxide; honey bees are prematurely aged by these. Therefore, the use of anaesthetics should, wherever possible, be avoided in ecological and behavioural studies.

Release

The release of the animals after marking is an operation that is too often casually undertaken. A few methods allow the animals to be marked without capture, others for it to be marked in the field and immediately released again. However, frequently after a period of incarceration, handling and disturbance, the animals are released into the field and it should not be surprising if they show a high level of activity immediately after release; indeed Greenslade recorded with individually marked ground beetles that there was far greater movement on the day after release than subsequently. Two approaches can be used to minimize this effect.

If the animal has a marked periodicity of movement (i.e. is strictly diurnal or nocturnal) then it should be released during its inactive period; for example I released radioactively tagged individuals of the frit fly *(Oscinella frit)* at dusk (its period of activity is from dawn to late afternoon). Animals that are active at most times of the day may be restrained from flying immediately after release by covering them with small cages. The release sites should be chosen carefully. It is especially important to avoid the release of small flying insects in the middle of the day when their escape flights may carry them beyond the shelter of the habitat into winds or thermals that can transport them for miles. Of course, only apparently healthy, unharmed individuals should be released.

The release points should be scattered throughout the habitat, as it is essential that the marked animals mix freely with the remainder of the population; e.g. Muir returned the arboreal mirids he had marked to all parts of the tree. Very sedentary animals may indeed invalidate the use of the capture-recature method if they do not move sufficiently to re-mix after marking, as Edwards found, surprisingly, with a population of the grasshopper, *Melanoplus*. The extent of the re-mixing may be checked, to some degree, by a comparison of the ratio of marked to unmarked individuals in samples from various parts of the habitat; the significance of the difference may be tested by a χ^2.

CAPTURE-RECAPTURE METHODS OF ESTIMATING POPULATION PARAMETERS

Lincoln Index Type Methods

Assumptions

Various assumptions underlie all methods of capture-recapture analysis. If the particular animal does not fulfil one or more conditions it might be possible to allow for this to some extent, but a method of analysis should not be applied without ensuring, as far as is practicable, that its inherent assumptions are satisfied.

The following assumptions underlie most methods of analysis:

(1) The marked animals are not affected (neither in behaviour nor life expectancy) by being marked and the marks will not be lost.

(2) The marked animals become completely mixed in the population.

Cold and inclement weather soon after release may seriously restrict the mixing of marked individuals in the population.

(3) The probability of capturing a marked animal is the same as that of capturing any member of the population; that is, the population is sampled randomly with respect to its mark status, age and sex. Termed 'equal catchability', this assumption has two aspects: firstly, that all individuals of the different age groups and of both sexes are sampled in the proportion in which they occur; secondly, that all the individuals are equally available for capture irrespective of their position in the habitat.

(4) Sampling must be at discrete time intervals and the actual time involved in taking the samples must be small in relation to the total time.

When using the simple Lincoln Index it is also assumed that:

(5) The population is a closed one or, if not, immigration and emigration can be measured or calculated.

(6) There are no births or deaths in the period between sampling or, if there are, allowance must be made for them.

Most of the other more complex methods of estimating population by capture -recapture may be applied in situations where either migration, natality or mortality or all three are occurring. These methods, however, require a series (at least two) of occasions on which animals are marked, on the second and subsequent occasions the recaptured animals are remarked and released again. Hence these methods make the further assumption:

(7) Being captured one or more times does not affect an animal's subsequent chance of capture. This is a further ex-

tension of assumption (3) – equal catchability.

Several of these methods estimate the number of marked animals available for recapture from their subsequent survival and that of other cohorts. These approaches therefore make the additional assumption:

(8) Every marked animal has the same probability of surviving through the sampling period. This means that if mark status is in any way related to age then mortality must be independent of age; it must act randomly on the population and some individuals (e.g. of greater age and different mark status) must not be at greater risk.

The Validity of the Assumptions

Marking has no effect

The experimental assessment of the effects of marking, the precautions in this respect and the re-mixing of the marked animals in the population have already been discussed above.

Manly has described a test to determine if the survival of individuals immediately after marking and release is different from that of animals that have been marked for some while. It is thus a test of the effect of the marking process and not of the influence of bearing a mark. These losses are referred to by fisheries workers as Type 1 losses. It depends on each animal bearing a unique mark and may be used in conjunction with the actual data derived from a series of, at least four, marking occasions. The difference (y) between the survival of animals newly exposed to marking and others is given by:

$$y = \log_e\left(\frac{r_i'}{r_i} \times \frac{u_i}{R_i}\right) \qquad (10.1)$$

where r'_1 = the number of marked animals in the ith sample that are recaptured again later, r_i = the total marked animals in the ith sample, u_i = the total unmarked animals in the ith sample and R; = the total animals marked on the ith occasion and subsequently recaptured (see Table 10.2). Now if the process of marking has no effect on the animal, y will have an expected value of zero.

The significance of any value of y can be determined by the statistic g, derived by dividing y by its variance:

$$g = y \Big/ \left(\frac{1}{r_i'} + \frac{1}{R_i} - \frac{1}{r_i} - \frac{1}{u_i} \right)^{\frac{1}{2}} \tag{10.2}$$

g will behave as a random normal variate with mean and variance zero. The

one-tail test is normally justified because an increase in survival due to the marking process can be ruled out and thus at the 5 % level $g < 1.64$ or at 10 level $g < 1.29$ may be taken as showing that the marking process has no effect on survival. A similar test has been developed by Robson.

Equal catchability

There may be several different causes for the violation of this assumption; underestimates of the population result. Roff has reviewed the statistical tests available for the purpose of testing for equal catchability but, after simulation, has concluded, that under practical conditions they may be unreliable. Such a warning is useful and means that the ecologist must also rely on other biological indications and on other methods for confirmation that his population is homogeneous: the computer cannot replace biological insight. It is to be hoped, however, that further refinements of the mathematical aspect will be achieved, for the excessive pessimism of Roff's concluding sentence does not advance the subject! Phillips & Campbell's studies on the whelk, suggest that 'frequency of capture' methods may be more robust and, as stressed above, the ecologist should always endeavour to select homogeneous categories.

(1) Sub-categories of the population are differently sampled because by the nature of the habitat only part of the population is available. This problem is particularly acute with subcortical, wood-boring and subterranean insects. It is well exemplified by ants: the number and composition of the foragers of an ant colony are so variable that unless one has a considerable knowledge of these aspects, or can sample from within the nest, Lincoln Index and related methods cannot be used for population estimation. The

marking and recapture of foraging ants may however be used to estimate the foraging area and, over a short period of time, the number of foragers:

$$F = R_f \times (T_0 + T_i) \qquad (10.3)$$

where F = the number of foragers; R_f = rate of flow of foragers per unit time, i.e. the average number of ants passing fixed points on all routes from a nest per minute; T_o = average time (in the same units as *RF)* spent outside the nest (found by marking individuals on the outward journey and timing their return); and T, = average time spent inside the nest (found by marking ants as they enter and timing their reappearance). In *Pogonomyrmex badius* only about 10 % of the total population forage above ground in any two-week period, whilst in the wood ant, *Formica,* a constant group of individuals forage such that if their number (F) is estimated, colony size (C) is given by:

$$\log \hat{C} = (\log.\ \hat{F} + 0.75)/1.01 \qquad (10.4)$$

Foragers of this species may remain away from the nest for up to two days.

(2) Sub-categories of the population are differentially sampled because differences due to sex and/or age or other causes. In many animals, especially when trapping techniques are used, differences will be associated with differences in sex or age. It is indeed a good principle to enumerate and estimate males and females separately, at least in the first instance; many workers have found striking differences in their behaviour. Initially different stages or age classes should also be considered separately and their survival rates determined using one of the formulae given below; if their survival rates differ significantly, one must continue to consider them separately. Indeed every attempt should be made to ensure that the population estimated is as homogeneous as possible.

If there is no mortality, a χ^2 goodness-of-fit test may be applied to a table of the releases, recaptures and non-recaptures of the different classes. Lomnicki found that land snails are very non-random in their movements, some individuals being more inclined to recapture than others.

(3) There is a periodicity in the availability of sub-categories.

This problem has been particularly encountered with mosquitoes, where an initial period of emigration and the gonotrophic and linked feedingcycles invalidate the equal catchability assumption and may lead to five fold over-estimates. Other animals show periodicities in their behaviour and therefore, particularly when sampling is by a behavioural method (e.g. trapping) and the periodicity cycle is long compared with the sampling interval, workers should consider this problem. With the Fisher & Ford method (see below) it may be allowed for by introducing additional terms into the model (see below). The modification of the Fisher & Ford method in this way depends on a knowledge of the periodicities gained independently. The extent to which the incorporation of these additional terms improves the relationship of the expected and observed recaptures gives a measure of their value, but a precise statistical test is not available.

(4) The processes of capturing and marking affect catchability. This can have two origins: the initial experience may alter catchability or it may be a cumulative effect. In the former case (more relevant to vertebrates) a regression method analogous to removal trapping may be used to estimate total population. The effect of repeated capture may be tested for by Leslie's test for random recaptures. A comparison is made of the actual and expected variances of a series of recaptures of individuals known to be alive throughout the sampling period; individual marks must have been used. This test is best illustrated by a worked example taken from Leslie's appendix, based on the recaptures of shearwaters; with insects, of course, the recapture periods would be days or weeks rather than years.

Example (after Leslie): 32 individuals marked in 1986 were recovered for the last time in 1992, therefore they were available for recapture in the years 1987-91 inclusive and the following two tables can be prepared:

Year by year analysis		All years analysis	
Year	No. of recaptures in each year (n_i)	No. of recaptures for each individual x	Frequency of x $f(x)$
1987	7	0	15

1988	7	1	7
1989	6	2	7
1990	4	3	2
1991	7	4	1
n_i =	31	5	0

$N = \Sigma f(x) = 32$

The actual sum of squares:

$$\Sigma(x-\bar{x})^2 \, \Sigma x^2 f(x) - \frac{[\Sigma xf(x)]^2}{N} \quad (10.5)$$

$$= 69 - 30.03 = \underline{38.97}$$

The expected (=theoretical) variance:

$$\sigma^2 = \frac{\Sigma n_1}{N} - \frac{\Sigma(n_i^2)}{N^2} \quad (10.6)$$

$$= \frac{31}{32} - \frac{199}{32^2} = 0.9688 - 0.1943 = \underline{0.7745}$$

Then $$X^2 = \frac{38.97}{0.7745} = \underline{50.32}$$

X^2 may be treated as equivalent to a $Y._2$ and for degrees of freedom $(N-1)$ of between 20 and 30 the probability of a value as great or greater than this can be assessed from χ^2 tables; probabilities of less than 0.50 imply that capture is not random. With values over 30, use can be made of the fact that $\sqrt{2\chi^2} - \sqrt{2\text{d.f.}-1}$ is approximately normally distributed about a mean of zero without standard deviation; in other words one calculates the value for this expression and looks up its probability in the table of the normal deviate. In the present case:

$$\sqrt{2 \times 50.32} - \sqrt{(2 \times 31) - 1} = 10.03 - 7.81 = +2.22$$

The probability of a deviate as great as this is somewhere between 0.025 and 0.020 and therefore the shearwaters were not collected randomly, Leslie has suggested that the test should only be used when the number of individuals is 20 or more and the number of occasions on which recapture was possible is at least

3. This test will not distinguish whether the higher catchability of some individuals is due to catching effects or, as Lomnicki presumed with land snails, inherent individual differences. Carothers has developed an extension of Leslie's test more appropriate when the number of occasions on which recapturing occurs is small.

Methods of Calculation

Although simple Lincoln Index estimates are not permissible unless conditions 5 and 6 are satisfied, the variants of the methods have all been derived to allow for loss (emigration and death) from or gains (immigration and birth) to the population. Migration rate can also be measured by other means or it may be artificially prevented by the use of a field cage. Birth and death can also be measured by other methods. Sometimes mortality rate may be so high that the interval between the release of the marked individuals and the recapture sampling is extremely short, a matter of hours; this is only possible with a mobile insect and under such conditions Craig's method might be used as well as one based on the Lincoln Index.

The Lincoln Index

Provided the first six conditions listed above are satisfied it is legitimate to estimate the total population from the simple index used by Lincoln:

Total population ÷ Original number marked = Total second sample ÷ Total recaptured

$$\hat{N} = \frac{an}{r} \tag{10.7}$$

where $\hat{N}$ = the estimate of the number of individuals in the population (N), n = total number of individuals in the second sample, a = total number marked and r = total recaptures.

Strictly the size of n should be predetermined and normally is approximately equal to a, then the variance of this estimate is given by:

$$\text{var}\,\hat{N} = \frac{a^2 n(n-r)}{r^3} \tag{10.8}$$

If the second sample *(n)* consists of a series of subsamples and a large proportion of the population have been marked then it is poss;ble to utilize the recovery ratios (= *r/n)* in each to calculate the standard error of the estimated population (Welch, 1960):

$$\hat{N} = \frac{a}{R_T} \qquad (10.9)$$

where *RT* = the recovery ratio *(r/n)* based on the total animals in all of the samples; the variance is approximately:

$$\text{var}\,\hat{N} = \left(\frac{a}{R_T{}^2}\right)^2 \times R_T \frac{(1-R_T)}{y} \qquad (10.10)$$

where y = total animals in subsamples. This approach is only valid if the

marked individuals are distributed randomly in the subsamples, which may be tested by comparing the observed and theoretical variance.

The above methods are applicable to large samples where the value of r is fairly large (say over 20); Bailey has suggested that with small samples a less biased estimate is given if 1 is added to ***n*** and *r*, i.e.:

$$\hat{N} = \frac{a(n+1)}{r+1} \qquad (10.11)$$

An approximate estimate of the variance of this is given by:

$$\text{var}\,\hat{N} = \frac{a^2(n+1)(n-r)}{(r+1)^2(r+2)} \qquad (10.12)$$

These methods are based on what is referred to as *direct sampling* in which the size of *n* is predetermined. If it is possible, *inverse sampling,* where the number of marked animals to be captured (i.e. r) is predetermined. has the advantages of giving an unbiased population estimate and variance (Bailey,

$$\hat{N} = \frac{n(a+1)}{r} - 1 \qquad (10.13)$$

$$\text{var}\,\hat{N} = \frac{(a-r+1)(a+1)n(n-r)}{r^2(r+1)} \tag{10.14}$$

A modification of the Lincoln Index has been develope by Gaskell & George. It is based on Bayes' Theorem that links the probability distribution of a new estimate with the likelihood of the sample and the previous estimate The incorporation of the previous estimate is of considerable practical significance: the value as a biological check of independent estimates by alternative methods is stressed throughout this book. This Bayesian modification now allows their incorporation into the final population estimate. Gaskell & George's formula is essentially:

$$\hat{N} = \frac{an+2N'}{r+2} \tag{10.15}$$

where N' = the previous estimate obtained, for example, from quadrat sampling or nearest neighbour type techniques.

Other Single Mark Methods

This approach is extensively used in fisheries where the ecologist can mark a number of fish, but must rely on commercial fishing to provide the recapture sample; one of the major problems in this work is to distinguish mortality due to man from that due to other causes.

If a group marking method is used, recaptured individuals must be removed and the successive removals recorded; with individual marks the number of previously un-recaptured individuals must be recorded: the fall off in these values, determined by regression, may be related to total population. The methods used will not be discussed further here; some are described by Beverton & Holt and the whole approach lucidly reviewed by Seber.

Review of Methods for a Series of Marking Occasions

If the animals are marked on a series of two or more occasions, then an allowance may be made for the loss of marked individuals between the time of the initial release and the time when the population is estimated. These methods have recently been reviewed by Cormack, Parr, Gaskell & George, Robson and Seber.

Their history and development, with particular reference to the entomological literature are given below. A simple, non-algebraic development of the main formulations is provided by Cormack, who shows how the estimates are based on the largest 'known groups'.

Jackson and Dowdeswell, Fisher & Ford were the first to devise methods for estimating the population using the data from a series of marking and recapture occasions. Besides assumptions 1-4 and 7 listed above, they also assumed a constant survival rate over a period of time; even though this was known to be an approximation it was considered necessary for an algebraic solution.

Jackson developed two methods in his work on the tsetse fly where he calculated a theoretical recapture immediately after release by either the 'positive method', where the loss ratio was calculated for that day over all release groups, and the 'negative method', where the loss ratio was calculated from the subsequent daily percentage losses for the two release groups. Richards & Waloff used Jackson's negative method and give worked examples. It is possible that the negative method might still be found useful in a situation, as provided by the use of radioactive isotopes, where marking is carried out on a limited number of occasions followed by a long series of recaptures.

Fisher's method is often referred to as the 'trellis method', as the data are initially set out. in a trellis diagram; details of its working are given by Dowdeswell. The survival rate has to be determined by trial and error, but with the advent of access to computers this comparatively robust method has enjoyed a revival. A useful discussion and summary of these early methods is given by MacLeod.

A slightly different method of analysis, but with the same principles, is given by Leslie & Chitty and Leslie and referred to as method A. It has been used with carabids by Griim and trypetid flies by Sonleitner & Bateman. Although these methods are of considerable historical interest and have been widely used, it is difficult to envisage a situation where their use would still be recommended in preference to the more recent methods. A possible exception to this statement is the use of Jackson's negative method under the circumstances outlined above.

An important advance was made by Bailey, who introduced maximum likelihood techniques into capture-recapture analysis and so was able to calculate the variances of his estimates. The equations in Bailey's triple-catch method are simple and may be solved directly to provide estimates of various population parameters.

Another significant step was made at about the same time by Leslie, who compared three different methods of classifying the animals according to their marks. In Leslie's method A, mentioned above, the animals were classified according to the occasion on which they were marked and thus an animal bearing several marks would give several entries in the recapture table. Jackson, Fisher and Bailey all also used this method of classification in drawing up their recapture tables, but Leslie & Chitty showed that it leads to loss of information. This loss is only slight when there are but a small number of multiple recaptures. His method B, in which the animals were classified according to the date on which they were last marked (ignoring all earlier marks), was completely efficient under the assumptions he made. Leslie gives a full account of his method and a worked example, but his formulae require solution by iterative methods when more than three sampling occasions are considered and hence the calculations are rather laborious. Jolly simplified the calculationss by providing formulae for explicit solutions (i.e. direct, not 'trial and error'). This method is particularly applicable if a fairly large number of individuals have been recaptured several times (multiple recaptures) in a long series of samples, a situation that often arises with work on mammals. However, Sonleitner & Batemanused Leslie's formulae to obtain estimates of the population of a trypetid fly and Muirused Jolly's method with a mirid bug.

Working on blowflies, where the incidence of recapture was very low, MacLeod found that none of the then known methods were usable and he derived two formulae (based on the Lincoln Index) which gave estimates, albeit rather approximate, of the total population. For both methods the mortality rate had to be ascertained independently, in his case in the laboratory, and for one method the percentage of flies not immigrating over a given time was measured by a separate experiment.

All the above methods are based on deterministic models that assume that the survival rate over an interval is an exact value, whereas it would be more correct to state that in nature an animal has a probability of surviving over the interval. This probability is well expressed by a stochastic model, but initially it was thought that the computations arising from a stochastic model would be too complex. Darroch showed that under certain conditions a fully stochastic model, giving explicit solutions for the estimation of population parameters, was possible, and Seber and Jolly have, independently, extended this method to cover situations in which there is both loss (death and emigration) and dilution (births and immigration). Their methods give similar solutions, except that Jolly's makes allowance for any animals killed after capture and hence not released again, a common occurrence in entomological experiments.

The Jolly-Seber method efficiently groups the data and is fully stochastic; however, its reliability strictly depends on assumption, effectively the probability of any animal surviving through any period is not affected by its age at the start of the period. Manly & Parr pointed out that with shortlived adult insects (e.g. many Lepidoptera) studied over their emergence period the animals marked on the ith occasion are likely, at that time, to be younger than those originally marked on the first. This in itself would not invalidate the Jolly-Seber method, if all deaths were independent of age: this cannot be assumed. They therefore devised a method free of assumption, based on intensity of sampling, but depending on a relatively high frequency of multiple recapture. The term 'multiple recapture' has been used extensively and is appropriate with mark-grouping methods where the number of marks an animal bears (or the number of occasions on which it has been captured) affects its classification. However, it will be noted that in the date-grouping methods no significance is attached to any mark other than the last and hence, as Jolly points out, the term multiple recapture should not be used in connection with these methods.

Choice of Method for a Series of Marking Occasions

Comparative studies on actual or simulated data have been made by several workers including Parr, Sheppard *et al.*, Manly, Roff and Bishop & Sheppard. Insights may often be gained if different methods are used and their discrepancies carefully considered. There are, however, certain general indications: the use and comparison of more than one method of estimation can be instructive.

As mentioned above the advent of ready access to computers has increased the utility of Fisher & Ford's trellis method: it is a robust method for small samples and, provided the survival rate remains relatively constant, gives population estimates not dissimilar to those of the Jolly-Seber method. It has the additional advantage that periodicity of availability for sampling, known from biological knowledge, may be incorporated in the trellis model.

Bailey's Triple Catch and the Jolly-Seber method both provide estimates of the variance, however, these have been shown to be related to the population estimate and are particularly unreliable for small samples. The Jolly-Seber method population estimate is usually reliable when 9% or more of the population is sampled and the survival rate is not less than 0.5; it is less sensitive to age-dependent variations in the mortality rate than Fisher & Ford's method, but is not of course independent of them. Except when there are only a very limited number of sampling occasions the Jolly-Seber method is superior to Bailey's.

The Jolly-Seber method may seriously overestimate the survival rate; it remains however probably the most useful method.

Manly & Parr's method is not affected by age-dependent mortality but required the sampling of a relatively high proportion of the population (>25 for populations under 250, > 10 % for larger populations). When these conditions can be achieved it should be used if mortality is thought to be related to age, particularly if it is high early in the life span studied.

The Fisher — Ford Method

The data from a series of recaptures are set out in a 'trellis diagram'; a constant survival rate that best fits the data is found by a trial and error process, and one essentially makes a series of Lincoln Index estimates working backwards in time from day to day from when the releases were originally made. Then the population estimate $\left(\hat{N}t\right)$ is:

$$\hat{N}_t = \frac{n_t a_i \phi_i}{r_{ti}} \tag{10.16}$$

where n_t = total sample at time t, a_i = total marked animals released at time i, ϕ_{i-t} = the survival rate over the period i – t and r_{ti}= the recaptures at time t of animals marked at time i. Details of the method are given by Fisher & Ford and in an expository form by Dowdeswell and Parr. Bishop & Sheppard provide a computer programme in ALGOL.

As already mentioned, one of the strengths of this robust model is that additional variables such as the periodicity of availability may be incorporated in the model and the computer used to find the best fit to the various parameters. A model for the mosquito, *Aedes aegypti,* assuming a four-day feeding (and hence availability) cycle with some secondary feeding on the second and third days, is shown in Table 10.2. The total recaptures for any day will have components bearing the marks of different days. For example, r, may contain recaptures from the first second and third marking occasions, and the values of these will, if the model is appropriate fit the expressions: $p_4 a_1 \phi^3 \beta_3, p_4 a_2 \phi^2 \beta_2$ and $p_4 a_3 \phi \beta_1$ respectively. The sampling intensity (catch rate), p, will vary from occasion to occasion; here it is the value for the fourth occasion. The marked animals (from a_i) available for recapture are modified by the survival rate, ϕ and availability proportion β. As it is a four day cycle β itself will have four values and each represents the different proportions available (in this case feeding) on each day of the cycle: the value of β_0 is 1 (all survivors are available on the fourth day), the other β-values, given by Conway *et al.* are somewhat complex algebraically because in their situation the periodicity of availability was complicated by some mosquitoes feeding on the second and third days of the cycle, as well as on the first.

However if we simplify and generalize the model and assume that any cohort of the animals in question is only available for sampling every fourth day, $\beta_0 = 1$ and β_1, β_2 and $\beta_3 = 0$. Certain terms will then disappear from the trellis (Table 10.2). The survival rate and β-values for the trellis may be determined by the normal least squares method and then the sampling intensities for each catch date determined.

In the simplified general case, with synchronized availability, the total population for the ith occasion is then estimated by

$$\hat{N}_i = \frac{n_i(1-\phi^x)}{p_i\phi^y(1-\phi)} \tag{11.17}$$

where x = the period of the cycle of availability (e.g. four days) and y = the period between birth and becoming available for capture (e.g. for mosquitoes one day). If y is zero, then the denominator becomes $p_i(1-\phi)$.

Full details for the special case of *Aedes aegypti* are given by Conway *et al.*; but the availability of a number of insects (including the rhinoceros beetle) to trapping methods shows a marked periodicity related to age, and the method is therefore of wider application.

Bailey's Triple-Catch Method

As indicated above this method is based on a deterministic model of survival and uses an inefficient method of grouping the data, although the latter is not a serious fault if very few animals are recaptured more than once. The formulae for the calculation of the various parameters are relatively simple. De Lury has suggested that a series of only three samples is unlikely to measure the magnitude of biological variation encountered. However birth- and death-rates may well be more constant over a short period than over a larger one and furthermore, as pointed out earlier if these (or any) estimates can be related to other estimates based

TABLE 10.2 : POPULATION MODEL FOR MARK AND RECAPTURE WITH CONSTANT SURVIVAL (Φ), BUT CYCLING PROPORTIONATE AVAILABILITY (Β) AND VARYING SAMPLING INTENSITY (CATCH RATE) (*P*).

1 n_l a_1							
2 n_2 a_2	$p_2 \alpha_1 \phi \beta_1$						
3 n_3 a_3	$p_3 \alpha_1 \phi^2 \beta_2$	$p_3 \alpha_2 \phi \beta_1$					
4 n_4 a_4	$p_4 \alpha_1 \phi^3 \beta_3$	$p_4 \alpha_2 \phi^2 \beta_2$	$p_4 \alpha_3 \phi \beta_1$				
5 n_5 a_5	$p_5 \alpha_1 \phi^4 \beta_0$	$p_5 \alpha_2 \phi^3 \beta_3$	$p_5 \alpha_3 \phi^2 \beta_2$	$p_5 \alpha_4 \phi \beta_1$			
6 n_6 a_6	$p_6 \alpha_1 \phi^5 \beta_1$	$p_6 \alpha_2 \phi^4 \beta_0$	$p_6 \alpha_3 \phi^3 \beta_3$	$p_6 \alpha_4 \phi^2 \beta_2$	$p_6 \alpha_5 \phi \beta_1$		
7 n_7 a_7	$p_7 \alpha_1 \phi^6 \beta_2$	$p_7 \alpha_2 \phi^5 \beta_1$	$p_7 \alpha_3 \phi^4 \beta_0$	$p_7 \alpha_4 \phi^3 \beta_3$	$p_7 \alpha_5 \phi^2 \beta_2$	$p_7 \alpha_6 \phi \beta_1$	
8 n_8 a_8	$p_8 \alpha_1 \phi^7 \beta_3$	$p_8 \alpha_2 \phi^6 \beta_2$	$p_8 \alpha_3 \phi^5 \beta_1$	$p_8 \alpha_4 \phi^4 \beta_4$	$p_8 \alpha_5 \phi^3 \beta_3$	$p_8 \alpha_6 \phi^2 \beta_2$	$p_8 \alpha_7 \phi \beta_1$

on different methods and assumptions then the reliability of the estimates is considerably strengthened. The mathematical background to this method is given by Bailey and it is described by Bailey, Richards and MacLeod. It was found particularly appropriate by Coulson in a study on craneflies, *Tipula,* with very short life expectancies, but series of triple-catch estimates may be used to estimate a population over a longer period. It seems that the results are most reliable when- large numbers are marked - and recaptured. For the purposes of illustration the sampling is said to take place on days 1, 2 and 3; the intervals between these sampling occasions may be of any length, provided they are long enough to allow for ample mixing of the marked individuals with the remainder of the population and not so long that a large proportion of the marked individuals have died.

With large samples the population on the second day is estimated:

$$\hat{N}_2 = \frac{a_2 n_2 r_{31}}{r_{21} r_{32}} \tag{11.18}$$

where a_2 = the number of newly marked animals released on the second day; n_2 = the total number of animals captured on the second day and r = recaptures with the first subscript representing the day of capture and the second the day of marking; thus r_{21} = the number of animals captured on the second day that had been marked on the first, r_{31} = the number of animals captured on the third day that had been marked on the first. It is clear that we are really concerned with the number of marks and hence the same animal could contribute to r_{31} and r_{32}. The logic of the above formula may be seen as follows:

$$\hat{N}_2 = \frac{\partial_1 n_2}{r_{21}} \tag{10.19}$$

which is the simple Lincoln Index with a_1 = the estimate of the number of individuals marked on day 1 that are available for recapture on day 2. Now if the death-rate is constant:

$$\frac{\hat{a}}{a_2} = \frac{r_{31}}{r_{32}}$$

$$\hat{a}_1 = \frac{a_2 r_{31}}{r_{32}} \tag{10.20}$$

Substituting for $\hat{a}_1$ in equation 10.19 above we arrive at the formula for the population given above. The large-sample variance of the estimate is:

$$\text{var}\ \hat{N}_2 = \hat{N}_2^2\left(\frac{1}{r_{21}} + \frac{1}{r_{32}} + \frac{1}{r_{31}} - \frac{1}{n_2}\right) \tag{10.21}$$

Where the numbers recaptured are fairly small there is some advantage in using 'Bailey's correction factor', i.e. the addition of 1 so that:

$$\hat{N}_2 = \frac{a_2(n_2+1)r_{31}}{(r_{21}+1)(r_{32}+1)} \tag{10.22}$$

with approximate variance:

$$\text{var}\ \hat{N}_2 = \hat{N}_2^2 - \frac{a_2^2(n_2+1)(n_2+2)r_{31}(r_{31}-1)}{(r_{21}+1)(r_{21}+2)(r_{32}+1)(r_{32}+2)} \tag{10.23}$$

The loss rate, which is compounded of the numbers actually dying and the numbers emigrating, is given by:

$$\underset{t=0\to1}{\gamma} = -\log_e\left(\frac{a_2 r_{31}}{a_1 r_{32}}\right)^{1/t_1} \tag{10.24}$$

where t_1 = the time interval between the first and second sampling occasions. The dilution rate, which is the result of births and immigration, is given by:

$$\underset{t=1\to2}{\beta} = \log_e\left(\frac{r_{21} n_3}{n_2 r_{31}}\right)^{1/t_2} \tag{10.25}$$

where t_2 = the time interval between the second and third sampling occasions.

Both these are measures of a rate per unit of time (the unit being the units of *t*); Bailey gives formulae for their variance.

Jolly-Seber stochastic method

The advantages of the stochastic model on which this method is based have already been discussed and its general utility has been indicated. The basic equation in Jolly's method is:

$$\hat{N}_i = \frac{\hat{M}_i n_i}{r_i} \tag{10.26}$$

where $\hat{N}_i$ = the estimate of population on day *i*, $\hat{M}_i$ = the estimate of the total number of marked animals in the population on day *i* (i.e. the counterpart of '*a*' in the simple Lincoln Index), *r;* = the total number of marked animals recaptured on day i and *n;* = the total number captured on day *i.*

The procedure may be demonstrated by the following example taken from Jolly. (The notation is slightly modified to conform with the rest of this section.)

(1) The field data are tabulated as in Table 10.3 according to the date of initial capture (or mark) and the date on which the animal was last captured. The columns are then summed to give the total number of (animals released on the *i*th occasion = *s*; of Jolly), subsequently recaptured (R_i), e.g. for day 7, R = 108.

(2) Another table is drawn up (Table 10.4) giving the total number of animals recaptured on day *i* bearing marks of day j or earlier (Jolly's a_{ij}); this is done by adding each row in Table 10.3 from left to right and entering the accumulated totals. The number marked before time i which are not caught in the ith sample, but are caught subsequently (Z_i), is found by adding all but the top entry (printed in bold) in each column. Thus Z_7 is given by the figures enclosed in Table 3.3. The figures above the line, i.e. the top entry in each column, represent the number of recaptures (r_i) for the day on its right, e.g. r_7 = 112.

(3) Then the estimate of the total number of marked animals at risk in the population on the sampling day may be made:

$$\hat{M}_i = \frac{a_i Z_i}{R_i} + r_i \tag{10.27}$$

TABLE 10.3 : THE TABULATION OF RECAPTURE DATA ACCORDING TO THE DATA ON WHICH THE ANIMAL WAS LAST CAUGHT FOR ANALYSIS BY JOLLY'S METHOD

Day of capture	**Total captured**	**Total released**	**Day when last captured (j)**												
i	n_i	a_i													
1	54	54	1												
2	146	143	10	2											
3	169	164	3	34	3										
4	209	202	5	18	33	4									
5	220	214	2	8	13	30	5								
6	209	207	2	4	8	20	43	6							
7	250	243	1	6	5	10	34	56	7						
8	176	175	0	4	0	3	14	19	46	8					
9	172	169	0	2	4	2	11	12	28	51	9				
10	127	126	0	0	1	2	3	5	17	22	34	10			
11	123	120	1	2	3	1	0	4	8	12	16	30	11		
12	120	120	0	1	3	1	1	2	7	4	11	16	26	12	
13	142		0	1	0	2	3	3	2	10	9	12	18	35	13
	$R_i =$			**80**	**70**	**71**	**109**	**101**	**108**	**99**	**70**	**58**	**44**	**35**	

Thus $$\hat{M}_7 = \frac{243 \times 110}{108} + 112 = 359.50$$

Similarly for other $\hat{M}_7$ the results being entered in Table 10.5 in which other population parameters are entered as calculated.

(4) The poportion of marked animals in the population at the moment of capture on day i is found and entered in the final table:

$$\alpha_i = \frac{r_i}{n_i} \tag{10.28}$$

$$\alpha_7 = \frac{112}{250} = 0.4480$$

(5) The total population is then estimated for each day (Table 10.5). (equations 10.26 and 10.28)

$$\hat{N}_i = \frac{\hat{M}_i}{\alpha_i}$$

(6) The probability that an animal alive at the moment of release of the ith sample will survive till the time of capture of the i + 1 th sample is found:

$$\hat{\phi}_i = \frac{M_{i+1}}{\hat{M}_i - r_i + a_i} \qquad (10.29)$$

TABLE 10.4 : CALCULATED TABLE OF THE TOTAL NUMBER OF MARKED ANIMALS RECAPTURED ON A GIVEN DAY BEARING MARKS OF DAY OR EARLIER

1						Day i-1						
10	*2*											
3	37	*3*										
5	23	**56**	*4*									
2	10	23	**53**	*5*								
2	6	14	34	**77**	*6*							
1	7	12	22	56	**112**	*7*						
0	4	4	7	21	40	**86**	*8*					
0	2	6	8	19	31	59	**110**	*9*				
0	0	1	3	6	11	28	50	**84**	*10*			
1	3	6	7	7	11	19	31	47	**77**	*11*		
0	1	4	5	6	8	15	19	30	46	**72**	*12*	
0	1	1	3	6	9	11	21	30	42	60	**95**	*13*

$Z(i-1)+1$

=	14	57	71	89	121	110	132	121	107	88	60
	Z_2	Z_3	Z_4	Z_5	Z_6	Z_7	Z_8	Z_9	Z_{10}	Z_{11}	Z_{12}

Survival rates estimates slightly over one may arise from sampling effects, but 'rates' greatly above this indicate a major error! Frequently it will be found that the marks of one occasion have been lost or were not recognized. This survival rate may be converted to a loss rate (the effect of death and emigration):

$$\hat{\gamma}_i = 1 - \hat{\phi}_i \tag{10.30}$$

(7) The number of new animals joining the population in the interval between the ith and i + lth samples and alive at time i + 1 is given by:

$$\hat{B}_i = \hat{N}_{i+1} - \hat{\phi}_i\left(\hat{N}_i - n_i + a_i\right) \tag{10.31}$$

This may be converted to the dilution rate (β):

$$\frac{1}{\beta} = 1 - \frac{\hat{B}_i}{\hat{N}_{i+1}} \tag{10.32}$$

(8) If desired the standard errors (the square roots of the variances) are obtained from:

$$\operatorname{var}\left(\hat{N}_1\right) = \hat{N}_i\left(\hat{N}_i - n_i\right)\left[\frac{\hat{M}_i - r_i + a_i}{\hat{M}_i}\left(\frac{1}{R_i} - \frac{1}{a_i}\right) + \frac{1+\alpha_i}{r_i}\right] + \hat{N}_i - \sum_{j=0}^{i=1}\frac{\hat{N}_i^2(j)}{\hat{B}_j} \tag{10.33}$$

The special problems involved in the computation of the summation term are discussed by Jolly (1965).

$$\operatorname{var}\left(\hat{\phi}_i\right) = \phi_i^2\left[\frac{\left(\hat{M}_{i+1} - r_{i+1}\right)\left(\hat{M}_{i+1} - r_{i+1} + a_{i+1}\right)}{\hat{M}_{i+1}{}^2}\left(\frac{1}{R_{i+1}} - \frac{1}{a_{i+1}}\right)\right.$$

$$\left.\frac{\hat{M}_i - r_i}{\hat{M}_i - r_i + a_i}\left(\frac{1}{R_i} - \frac{1}{a_i}\right) + \frac{1-\phi_i}{\hat{M}_{i+1}}\right] \tag{10.34}$$

$$\operatorname{var}\left(\hat{\beta}_i\right) = \frac{\hat{\beta}_i^2\left(\hat{M}_{i+1} - r_{i+1}\right)\left(\hat{M}_{i+1} - r_{i+1} + a_{i+1}\right)}{\hat{M}_{i+1}{}^2}\left(\frac{1}{R_{i+1}} - \frac{1}{a_{i+1}}\right)$$

$$\times\left[\frac{\phi_i a_i(1-\alpha_i)}{\alpha_i}\right]^2\left(\frac{1}{R_i} - \frac{1}{a_i}\right)$$

$$+\frac{\left(\hat{N}_i - n_i\right)\hat{N}_{i+1} - \hat{B}_i)(1-\alpha_i)(1-\phi_i)}{\hat{M}_i - r_i + a_i}$$

$$+\hat{N}_{i+1}\left(\hat{N}_{i+1} - n_{i+1}\right)\frac{1-\alpha_{i+1}}{r_{i+1}} + \phi_1^2\hat{N}_i\left(\hat{N}_i - n_i\right)\frac{1-\alpha_i}{r_i} \tag{10.35}$$

Jolly (1965) also gives terms for the covariance. These equations for the variance are fairly complex and if standard errors are required for a long series of estimates, then they should if possible be programmed on a computer.

(9) Jolly shows that the variances of the population and survival rate estimates given above contain an error component due to the real variation in population numbers, apart from errors of estimation. If only errors of estimation are required the formulae are:

$$\mathrm{var}\left(\hat{N}_i / N_i\right) = \hat{N}_i\left(\hat{N}_i - n_i\right)\left\{\frac{\hat{M}_i - r_i + a_i}{\hat{M}_i}\left(\frac{1}{R_i} - \frac{1}{a_i}\right) + \frac{1-\alpha_i}{r_i}\right\} \tag{10.36}$$

$$\mathrm{var}\left(\hat{\phi}_i | \phi_i\right) = \mathrm{var}\left(\hat{\phi}_i\right) - \frac{\hat{\phi}_i^2(1-\phi_i)}{\hat{M}_{i+1}} \tag{10.37}$$

The errors that arise in this method may be thought of as having two causes: (a) 'small sample biases' inherent in the method and (b) biases due to the violation of the assumptions. The former will be minimal when the number of recaptures is large, and therefore Jolly suggests that when the cost of marking is high, it might be an advantage if the sampling for recaptures was separated from the sampling and marking of further animals to provide the next value of a_i. Sampling for recaptures could be carried out extensively by less experienced staff than are needed to mark.

Manly and Roff have shown that these errors often express themselves as underestimates of the variance, because of the correlation between the estimates and their variances. Some improvement is obtained if N and ϕ are transformed to logarithms before making the calculations.

TABLE 10.5 : THE FINAL TABLE FOR A JOLLY TYPE MARK AND RECAPTURE ANALYSIS

Day	Proportion of	No. Marked animals	Total population	survival rates	No. of new	Standard errors			Standard errors due to errors the estimation of parameters itself	
i	$\hat{\alpha}_i$	$\hat{M}_i$	$\hat{N}_i$	$\hat{\phi}_i$	$\hat{\beta}_i$	$\sqrt{\{V(\hat{N}_i)\}}$	$\sqrt{\{V(\hat{\phi}_i)\}}$	$\sqrt{\{V(\hat{\beta}_i)\}}$	$\sqrt{\{V(\hat{N}_i / N_i)\}}$	$\sqrt{\left\{V(\hat{\phi}_i) - \frac{\hat{\phi}_i^2(1-\hat{\phi}_i)}{\hat{M}_{i+1}}\right\}}$
1	—	0	—	0.649	—	—	0.114	—	—	0.093
2	0.0685	35.02	511.2	1.015	263.2	151.2	.110	179.2	150.8	.110
3	.2189	170.54	779.1	0.867	291.8	129.3	.107	137.7	128.9	.105
4	.2679	258.00	963.0	.564	406.4	140.9	.064	120.2	140.3	.059
5	.2409	227.73	945.3	.836	96.9	125.5	.075	111.4	124.3	.073
6	.3684	324.99	882.2	.790	107.0	96.1	.070	74.8	94.4	.068
7	.4480	359.50	802.5	.651	135.7	74.8	.056	55.6	72.4	.052
8	.4886	319.33	653.6	.985	-13.8	61.7	.093	52.5	58.9	.093
9	.6395	402.13	628.8	.686	49.0	61.9	.080	34.2	59.1	.077
10	.6614	316.45	478.5	.884	84.1	51.8	.120	40.2	48.9	.118
11	.6260	317.00	506.4	.771	74.5	65.8	.128	41.1	63.7	.126
12	.6000	277.71	462.8	70.2	68.4 13	.6690				

The effect of unequal catchability on the estimates has been investigated by Carothers and Gilbert. Their studies show that reasonable estimates may be obtained if certain additional conditions cap be met. For example, if the population is closed and probabilities of capture remain constant, the mid-day of the sampling period will provide a less biased estimate; additional sampling days greatly reduce the bias; and provided all animals have a probability of capture > 0.5, few errors arise. Such a high probability of recapture is unusual but not impossible particularly in small populations [e.g. Southwood & Scudder's study of lacebugs] and it is in small populations that errors are particularly serious. Gilbert stresses that if this high probability of recapture can be achieved then unequal (heterogeneous) catchability is not in itself serious: for these special conditions, at least, a welcome counterblast to Roff's pessimism! Carothers shows that violation of the equal catchability assumption is not so serious if the variations in the probability of capture are symmetrical both in regard to time and the individuals: however, for populations with, for example, a high degree of 'trapshyness' the problem is more severe. This emphasizes the importance of ensuring that the collecting method and/or the collectors skill does not vary during the experiment; if, for example, insects were being collected by sweeping a change in collector could significantly affect the probability of individuals living near the base of the vegetation being recaptured.

The Jolly-Seber method has been used with a number of different insects. Computer programmes are provided by Davies and White.

Manly and Parr's Method

The assumption that survival probability is equal is often quite unjustified in entomological studies. To overcome this Manly & Parr, devised a method based on the intensity of sampling, each animal being considered to have the same chance of capture on the ith occasion, so that

$$\hat{N}_i = \frac{n_i}{p_i} \tag{10.38}$$

where p_i = the sampling intensity. In practice:

$$\hat{p}_i = \frac{r_i}{\hat{M}_i} \quad (10.39)$$

TABLE 10.6: MANLY & PARR'S METHOD OF RECORDING DATA FOR THEIR METHOD OF ESTIMATION (OBVIOUSLY ONE WOULD NORMALLY MARK FAR MORE THAN THREE ANIMALS ON EACH OCCASION).

	Sampling Days					
	1	**2**	**3**	**4**	**5**	**6**
1	*x*	*y*	*z*	*x*	—	—
2	*x*	*z*	*y*	*z*	*x*	—
3	*x*	—	—	—	—	—
4		*x*	*z*	*x*	—	—
5		*x*	*y*	*z*	*y*	*x*
6		*x*	*z*	*y*	*x*	—
7			*x*	*y*	*z*	*x*
8			x	z	x	—
9			*x*	*y*	*z*	*x*
$\Sigma y =$		1	2	3	1	
$\Sigma z_i =$		1	3	3	2	

where, as in the Jolly-Seber method, *M;* is the estimate of the total number of marked animals present in the population on the ith occasion: a summation of the survivors from the various *a*'s, so $\hat{M}_i \approx \hat{a}$ (as used in Bailey's Triple Catch). Thus Manly & Parr's equation becomes (equation 10.7):

$$\hat{N}_i = \frac{an_i}{r_i}$$

the basic Lincoln-Petersen index. The animals must be individually marked or have date-specific marks. Manly & Parr prepared an individual animal table allocating the symbol x for the first or last occurrence of an individual mark, y for the intermediate occasion when individual mark was captured and z for occasions when it was there, but not recaptured, i.e. the blanks left between the two x's after the y's have been inserted. A greatly curtailed example is given in Table 10.6. The sampling intensity will clearly be the ratio of those animals known to be present before and after, captured out of the total number available known to be present before and after. For any day these are given by Σy_i and $(\Sigma y_i + \Sigma z_i)$ respectively. Then:

$$\hat{p}_i = \sum y_i / \left(\sum y_i + \sum z_i\right) \qquad (10.40)$$

or (from Table 11.6)

$$\hat{p}_3 = 2/(2 + 3) = 0.4$$

Thus the population estimate is (equations 10.38 and 10.40):

$$\hat{N}_i = \frac{n_i}{p_i} = \frac{n_i(\Sigma y_i + \Sigma z_i)}{\Sigma y_i} \qquad (10.41)$$

survival may be estimated:

$$\phi_{i-(i+1)} = \frac{r_{i,i+1}}{n_i \times b_{i+1}} \qquad (10.42)$$

where $r_{i,i+1}$ = the animals caught in both the ith and (i + 1) samples.

The births, number of new animals entering the population, are estimated by:

$$\hat{B}_{i-(i+1)} = \hat{N}_{i+1} - \hat{\phi}_i \hat{N}_i \qquad (10.43)$$

Manly has developed an expression for the variance of the population estimate:

$$\text{var}\left(\hat{N}_i\right) = \frac{\hat{N}_i\left(\hat{N}_i - n_i\right)\left(\hat{N}_i - C_i\right)}{n_i C_i} \qquad (10.44)$$

where $C_i = \Sigma y_i + \Sigma z_i$

As mentioned this method is not robust unless a fairly large sample is taken. Seber shows how the necessary data may be tabulated in a form similar to that of the Jolly-Seber method: this would undoubtedly be more convenient with large sets of data, but Manly & Parr's original arrangement is simpler for exposition.

Frequency of capture methods (Schnabel census)

In considering the assumption of equal probability of capture it was necessary to examine the frequency of capture, many models and tests have been developed of which a few are referred to above. The information may be considered another way: if the frequency of capture follows a particular distribution (f_1 . . . animals captured once, f_2 . . . animals captured twice, f_3 ... f_x), then the term f_0 represents those animals that have not been marked or captured at all and the sum of all the terms will represent the total population. Theoretically problems do not arise from unequal catchability, a non-Poisson distribution may be fitted that describes the deviations from random in the probability of capture. Tanton fitted a zero truncated negative binomial to the frequency of recapture of wood mice, *Apodemus*. Further details of such models, particularly useful in small mammal work are given by Seber.

Craig's method: constant probability of capture

This method, devised for use with butterflies assumes that the population is closed (there are no births or deaths and the animal stays within its habitat) and there is a constant probability of capture. In practice the butterflies, or other highly mobile but colonial animals, are collected randomly, marked and immediately released, after recording the number of times, if any, the animal had previously been captured. The effective sample size is thus one. Craig assumed that the frequency of recapture could be described by two mathematical models: the truncated Poisson and the Stevens' distribution function. On the basis of these two models six different methods can be used to estimate the size of the population, three based on moments and three on maximum likelihood: Craig found, however, that they gave similar results. The two simplest are the moment estimates based on the Poisson; they are:

(1) $$\hat{N} = \left(\Sigma x f_x^2 / \left(\Sigma x^2 f_x - \Sigma x f_x\right)\right) \tag{10.45}$$

where $\hat{N}$ = the estimate of population, x = the number of times an individual had been marked, f_x = the frequency with which individuals marked x times had been caught and thus Exf_x = the total number of different times animals were captured (viz. 1 × the number caught once +2 × the number caught twice + 3 × the number caught thrice, etc). This method is simple, allowing a direct solution, but is subject to greater sampling error. It is useful for obtaining a trial value for use in solving the equation in method 2 (and in the other methods given by Craig).

(2) $$\log \hat{N} - \log\left(\hat{N} - \Sigma f_x\right) = \Sigma x f_x / N \tag{10.46}$$

Where the symbols are as above, Ef_x being the total different individual animals caught (viz. the number caught once + the number caught twice + the number caught thrice, etc.). Craig (1953) gives formulae for the variances of these estimates:

for method 1 $$\text{var}\,\hat{N} = \frac{2N}{\lambda^2} \tag{10.47}$$

where $\lambda = \Sigma x f_x$

for method 2 $$\text{var}\,\hat{N} = \frac{N}{\left(e^{\lambda} - 1 - \lambda\right)} \tag{10.48}$$

where e is the base of natural (Napierian) logarithms and thus the value of e^{λ} is found by using a table of natural logs 'backwards'.

It is clear that this method is based on different assumptions from the Lincoln Index. It demands that the animals be very mobile so that their chances of recapture are virtually random almost immediately after release and yet they must not leave the habitat. It has been used with butterflies and might be applied to other large conspicuous flying or very mobile animals under certain circumstances. Phillips & Campbell showed that the Schnabel Census method is more robust when portions of the population are inaccessible.

Chaage in ratio methods (Kelker's selective removal)

This method uses the natural marks of a population, normally the difference between the sexes, but theoretically any other recognizable distinction could be used (e.g. the different morphs of a polymorphic species). The proportion of the different forms or components are determined, a known number of one form is removed ('selective removal') and the new ratio found; from the change in the ratio the total population can be estimated:

$$\hat{N} = K_2 \div \left[D_{\alpha_1} - \frac{\left(D_{\beta_1} D_{\alpha_1}\right)}{D_{\beta_2}} \right] \quad (10.49)$$

where α and β are the components (e.g. sexes) in a population, a is the component part of which is removed during the interval between times 1 and 2, and fi is the component (or components) the member which are not removed, then K_α = number of component a that are killed; D = the proportion of the population represented by a component at time 1 or 2; thus D_{α_1} = the proportion of a in the population as a decimal before any were removed; D_{β_1} = the proportion of the population as a decimal represented by $ after K individuals of a were removed.

Such an approach is clearly most applicable with game where individuals of certain sex or age class are killed; it is some time referred to as the dichotomy method.

The mathematical background has been developed by Chapman, who compared the results from this method with those given by the Lincoln type capture -recapture analysis; he showed that capture -recapture estimation procedure will yield more information for the same amount of effort. A further objection to the use of this method with insect populations is that the selective removal and the resulting atypical unbalance could seriously prejudice any further study of the population; in populations of game animals such removal is one of the 'normal' mortality factors.

This method does not depend on the assumption that initial capture does not alter the probability of subsequent recapture, and this is an advantage over those methods that utilize artifical marks and might indicate conditions for its use, perhaps only as a test of estimates derived by Lincoln type methods. Chapman gives a technique for combining the two approaches. Paulik & Robson and Seber provide comprehensive reviews of this approach.

MECHANICAL METHODS OF EXTRACTION

Mechanical methods have the advantages that theoretically they extract all stages, mobile and sedentary, and are in no way dependent on the behaviour of the animal or the condition of the substrate; samples for mechanical extraction may be stored frozen for long periods before use. Their disadvantages are that compared with behavioural methods the operator must expend a great deal of time and energy on each sample, that sometimes they damage the animals and that, as mobile and immobile animals are extracted, it may be difficult to distinguish animals that were dead at the time of sampling from those that were ative.

There are a number of distinct mechanical processes that can be used to separate animals from the soil and vegetable material: sieving, flotation, sedimentation, elutriation and differential wetting. The following account, however, is intended to be functional, rather than classificatory or historical; only the main types of processes will be outlined and it must be stressed that although there are already many different combination and variants others will need to be developed for particular animals and substrates.

Dry Sieving

This method may be of use for the separation of fairly large (and occasionally small) animals, from friable soil or fallen leaves. Its disadvantages are that small specimens are often lost and a considerable amount of time needs to be spent in hand-sorting the sieved material. The Reitter sifter is a simple device, but this is really more a collector's tool than a means of estimating populations. Lane & Shirck's was the first mechanical sieve with a to-and-fro motion although it was worked by hand. The most complex apparatus of this type is the fully motorized self-propelled model of Lange, Akesson & Carlson, which was used for surveying elaterid larvae in California.

Molluscs have been extracted by dry sieving methods, but these have been shown to be inaccurate, some snails remaining in the leaves.

Dry sieving has also been used to separate mosquito eggs from mud. A modified grain drier is the basis of this method; the samples are dried until almost dusty, passed through a mesh sieve and then into the top of four shaker sieves of the grain drier. Coarser particles are removed by the first three sieves, but the eggs and similar sized soil particles are held on the last one (80-mesh/in); they then fall through an opening on to a 60-mesh in roll screen; a carefully adjusted air current blows away the lighter particles during the fall. Because of their spindle shape the eggs, with only a very small quantity of soil, eventually pass 'end-on' through the roll screen into the `catch pan'. Such a degree of mechanization is possible because only a single organism with a constant and regular shape and a uniform weight was required.

The active red-legged earth mite *(Halotydeus destructor)* lives amongst the grass and plant debris in Australian pastures; Wallace has shown that it can be very accurately sampled by taking shallow cores which are then inverted and rotated, still within the corer, over a sieve in a funnel. The sides of the metal corer are tapped a number of times and the mites fall through the sieve and may be collected in tubes below the funnel.

Soil Washing (or wet sieving)

This technique is most often used in conjunction with other methods (see below); on its own it is of particular value where the organisms are much smaller than the particles of the substrate (e.g. small snails amongst freshly fallen leaves) or much larger (e.g.many mud-dwellers), and hence separation by size alone is sufficient. The sieves may be of basic designs: flat, revolving or 'three dimensional'.

Flat sieves may be built up into a tower and a series of sieves was the basis of early uses of this method by N. A, Cobb and H. M. Morris. Sieving is often used in studies on the bottom animals of aquatic habitats for the separation of the mud from the organisms and stones that are retained by the sieve; a single frame is used into which frames with different mesh phosphor-bronze screens may be fitted. Jonasson has shown that many benthic organisms pass through the 0.6-mm mesh gauge sieves frequently employed in such studies; with insect larvae the size of the head capsule determined whether or not it would pass through the mesh of a sieve and there was no evidence that a significant proportion would roll up and so be retained by coarse sieves. The smallest sieve used by Jonasson had a mesh gauze of 0.2 mm, but clearly in any particular study its size would be determined by the diameter of the organism being sampled.

Although most soil nematodes can be separated elutriation or the Baermann funnel for those in marine habitats these methods may not give good results and sieving is recommended to separate the largest worms, followed by agitation, to make a suspension of the remainder, which is subsampled volumetrically. The subsamples are examined directly.

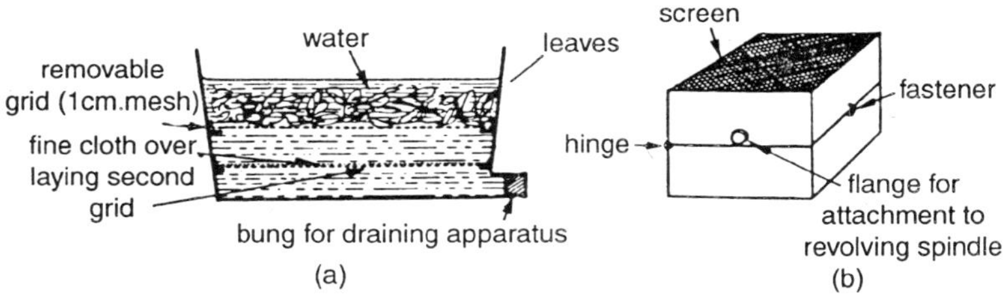

Figure 10.4 : ***a. Tank for the separation of molluscs from fallen leaves by wet sieving. b. Simple sieve box for wet sieving sawfly cocoons.***

Gastropod molluscs and possibly other small animals may be efficiently separated from fresh leaf litter by wet sieving, but here it is the organisms that pass through the sieve and the unwanted material that is retained. The leaves are placed on a coarse sieve (mesh size 1 cm), above a fine cloth-covered grid, the whole immersed in a vessel of water and the leaves are stirred from time to time; after about 15 minutes the vessel can be drained from below and the molluscs will be found on the cloth sheet, which is then inspected under the microscope. Clearly this method is not satisfactory for molluscs amongst soil or humus as much of these materials will pass through the upper sieve.

Simple revolving sieves were used by McLeod for the extraction of sawfly cocoons. They consisted of wooden boxes, hinged in the middle and with screen tops and bottoms; they were revolved once or twice while a jet of water from a hose was played on to them. More elaborate revolving models were designed by Horsfall and Read; these are discussed below.

The so-called 'three-dimensional' or rod sieve was designed by Stewart to facilitate the separation, undamaged, of tipulid larvae or other large animals from grassy soil samples, prior to flotation. Stewart's apparatus consists of thirteen tiers of parallel galvanized steel rods set in an open-ended box. The rods are progressively closer in each tier (2.5 cm closing to 1.3 cm apart); the side of the box may be removed (for cleaning, a process assisted by a sliding panel). The sample is placed on the top row of rods and the box moved up and down in a tank of water; the inside of the tank is lined with a normal screen sieve. The vegetation will pass dowri between the rods, freeing insects and mineral material which remain in the tank. The three dimensional sieve is removed (and if necessary cleaned) and the sieve inside the tank removed and the normal flotation procedure followed.

Soil Washing and Flotation

When the organisms to be extracted and the rest of the sample are of different particle-size, sieving alone is sufficient for extraction; however, in most situations after sieving the animals required remain mixed with a mass of similar sized mineral and vegetable matter. As the specific gravity of mineral and biotic material is frequently different the extraction may be taken a stage further by flotation of the animals; unfortunately plant material usually floats equally well. The combination of flotation, devised by A. Berlese many years before, with soil washing is usually associated with Ladell. This approach, much modified and improved by Salt & Hollick and Raw, is the basis of one of the most widely used, versatile and efficient techniques; it consists of four stages, three of which will be discussed under this heading.

(1) *Pretreatment.* In order to disperse the soil particles, particularly important if there is a high clay content, the cores may be soaked in water and deep frozen. Chemical dispersion may also be used: the core soaked in solutions of sodium citrate or sodium oxalate. For the heavy clay soils it may be necessary to combine chemical and physical methods: the core is gently crumbled into a plastic container and covered with a solution of sodium hexametaphosphate (50 g) and sodium carbonate (20 g) in one litre of water (sold commercially as 'Calgon' or 'Sparkaleen'); the whole is then placed under reduced pressure in a vacuum desiccator for a time. After the restoration of atmospheric pressure the sample is frozen for at least 48 hours and may be stored in this condition.

(2) *Soil washing.* The sample is placed in the upper and coarsest, sieve of the washing apparatus; it may be washed through by slow jets of water or single sieves 'dunked' in the settling can. The material retained in the sieves must be carefully teased apart, and thoroughly washed; a few large animals may be removed at this stage. The mesh of the lowest sieve is such as to allow the animals required to pass through into the settling can, which is pivoted and is then tipped into the 'Ladell can'. A 'three-dimensional' sieve may be necessary for the above process if there is a lot of grass in the sample and large animals are being separated. The Ladell, which resembles an inverted bottomless paraffin can, has a fine phosphor-bronze sieve and its lower opening immersed in the drainage tank; this tank should be arranged so that the level of water maintained in it, and hence in the Ladell, is slightly above the sieve of the latter-this minimizes blockage of the sieve. The Ladell is allowed to drain; this, often tedious, process may be aided by tapping the side of the can with the hand from time to time. The standard Ladell has a fine sieve (mesh size about 0.2 mm); for many animals this may be much finer than is necessary and to overcome this Stephenson has designed a Ladell, with a set of interchangeable sieve plates of various mesh sizes.

In conclusion it must be warned that soil washing is an invariably wet operation and is best carried out in a room with a concrete floor and the operator wearing rubber boots and suitable apparel.

(3) *Flotation.* The Ladell is removed from the tank to a stand, allowed to drain completely and then the process of flotation is begun. The lower opening of the Ladell is closed with a bung and the flotation liquid introduced until the Ladell is about two-thirds full. Concentrated magnesium sulphate solution (specific gravity c. 1.2) is the most usual flotation liquid, but solutions of sodium chloride, potassium bromide or zinc chloride may be used. Air is then bubbled up from the bottom of the Ladell; this agitation is continued for 2 - 3 minutes and serves to free any animal matter that may have been trapped on the sieve. More of the flotation solution is introduced from below until the liquid and 'float' passes over the lip and the latter is retained on the collecting tube, which usually consists of a glass tube with a piece of bolting silk held in place with a rubber band. A wash bottle may be used to help direct the float out of the Ladell. When the surface and sides of the Ladell are completely clean, the animal and plant material will all be in the collecting tube and the process is complete. The flotation liquid is then drained from the Ladell, for further use, by appropriate manipulations of the reservoir and pinchclips.

The animals may now be separated from the plant material by direct examination under the microscope or it may be necessary to carry out a further separation based on differential wetting and/or centrifuging. However, before these are discussed other washing and flotation techniques must be reviewed.

The apparatus of d'Aguilar *et al.* first removes the smallest silt particles by agitating the sample with a water current in a cylinder with a fine gauze side; subsequently the material is passed through a series of sieves, followed by flotation. Edwards, Whiting & Heath have developed a fully mechanised soil washing process that in general is as efficient as the Salt Hollick process described above.

The eggs and puparia of the cabbage root fly may be separated from the soil by sieving and flotation in water; with such insects therefore there is no need for the Ladell - the final sieve should

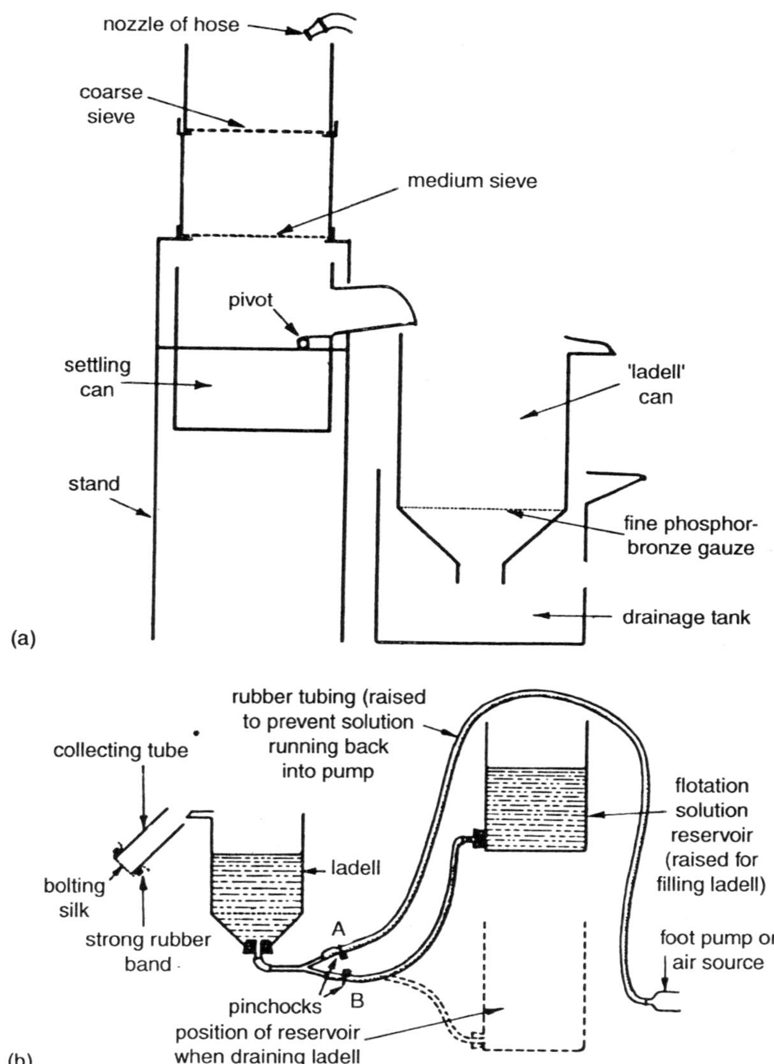

Figure 10.5 : ***a. Soil washing apparatus. b. Ladell can and associated equipment during the air agitation phase of flotation.***

be fine enough to retain them; this is immersed in water and the insects float to the surface. A certain amount of foam often develops in soil washing and hinders examination of the float; it may be dispersed with a small quantity of caprylic acid. To separate relatively large insects (root maggots) Read used a cylindrical aluminium screen sieve that was sprayed with water and rotated and half submerged in a tank; all the fine soil passed out of the sieve, which could then be opened and the floating insects removed. Gerard found a combination of wet sieving and simple magnesium sulphate flotation efficient for earthworm sampling, although Nordstrom & Rundgren's studies suggest that the additional labour involved, compared with chemical extraction, is not always justified.

A revolving sieve was also used by Horsfall for mosquito eggs. Here the function of the drum, which consisted of three concentric sieves, was to disintegrate the sample and retain the larger materials, the eggs passing out and being collected in the finest of a series of sieves.

Flotation

If the sample is first dried flotation alone may be used for molluscs and nematode cysts. Vagvolgyi's method has been modified by Mason who found it 84 % efficient. The sample is put into hot water and stirred, dead shells and litter float and are removed; badly broken shells, soil and the originally live snails, now killed by heat, sink. The sediment is then dried at 120°C for about 16 hours when the contents of the originally live snails contract; thus when the sample is carefully crumbled into a weak solution of detergent these now float and may be removed by careful searching. Badly broken shells, soil and other debris again sink. With nematode cysts drying, dry sieving and flotation in water is the usual sequence; various details and modifications are given by Goodey and in papers in Kevan and Murphy.

Often it may not be necessary to undertake the elaborate soil washing and flotation as described above. A very large range of benthic marine animals are efficiently extracted from grab-samples by merely stirring in a glass beaker with carbon tetrachloride. The mineral matter remains as a sediment, but the animals float. This technique would seem to have many applications for the extraction of invertebrates from freshwater and terrestrial substrates, so long as there is not a great quantity of plant matter. Indeed there is a range of heavy organic solvents whose potentialities have been little explored for ecological work; but care should be taken over ventilation when working with them. Sellmer used a 75 % solution of zinc chloride (sp. gravity 2.1) to float *Gemma,* a small bivalve mollusc, from marine sediment. Laurence separated insect larvae from dung by merely stirring the samples in 25% solution of magnesium sulphate, and Iversen obtained 75-92% of mosquito eggs for temporary woodland pools by this method, though Service added centrifugal flotation (see below). The efficiencies of this method, for the extraction of all stages of the midge *Leptoconops* from sand, were determined by Davies & Linley as: eggs 18 j; 1st instar larvae 29 %; 2nd instar larvae 72 %; 3rd & 4th instar larvae 95 % and pupae 85 %. Slugs can be separated by wet sieving and flotation and sodium chloride (salt) solution or brine has been widely used; calcium chloride for larval and pupal midges and sucrose solution (S.G.1.12) has been found effective for stream animals.

An interesting flotation technique using hydrogen peroxide is described by King for the separation of the eggs of the sugar-cane froghopper, *Aeneolamia varia.* After rotary sieving, the mixture of eggs, fine plant debris and sand was added to water containing one part per hundred of 6 hydrogen peroxide. Oxygen, produced from the hydrogen peroxide when it comes into contact with the plant debris, causes this to float to the surface. After its removal the eggs may be floated off in magnesium sulphate and glycerine solution (S.G.1.2).

The separation of plant and animal matter by differential wetting

A mixture of animal and plant material results from soil washing and flotation and also from various methods of sampling animals on vegetation.

There are three possible approaches to their separation, one depending on the flotation effect of oxygen derived from decomposing hydrogen peroxide and two dependent on the lipoid and waterproof cuticle of arthropods. Either the arthropod cuticle can be wetted by a hydrocarbon 'oil' or the plant material can be waterlogged so that it will sink in aqueous solutions.

(1) *Wetting the arthropod cuticle.* If arthropods and plant material are shaken up in a mixture of petrol or other hydrocarbon and water and then allowed to settle, the arthropods, whose cuticles are wetted by the petrol, will lie in the pertrol layer above the water and the plant material in the water. As Murphy has pointed out this separation will be imperfect if the plant material contains much air or if the specific gravities of the two phases are either too dissimilar or too close. The former may be overcome by boiling the suspension in water first; the second by the choice of appropriate solutions. 60 % ethyl alcohol may be used for the aqueous phase with a light oil. This concept was originally introduced into soil faunal studies by Salt & Hollick. The float from the collection tube is placed in a wide-necked vessel with a little water, benzene is added, the whole shaken vigorously and allowed to settle. The arthropods will be in the benzene phase, which may be washed over into an outer vessel by adding further water below the surface from a pipette or wash bottle. Raw introduced the idea of freezing the benzene (m.p. 50°C); the plug plus animals may then be removed and the bnenzene evatporated in a sintered glass crucible. (The fumes should be carried away by an exhaust fan as they are an accumulative poison.) A combination of decahydronophalene ('Dekalin') + carbon tetrachloride (S.G.1.2) and zinc sulphate solution (S.G.1.3) has been found particularly effective, the arthropods being floated off in the organic solvent mixture and the organic debris in the zinc sulphate solution run off in a separating funnel. Xylene and paraffin have also been used for the oil phase and gelatine, that solidifies on cooling, for the aqueous phase; in the latter case, it is the plant material that is removed in a plug. If the cuticle is not easily wetted by the benzene, extraction is improved if a solution with a specific gravity of about 1.3 is used instead of water and the initial mixture exposed to a negative pressure.

(2) *Waterlogging the plant material.* This process, by boiling under reduced pressure or vacuum extraction, may be used as part of the above technique. Repeated freezing and thawing also serves to impregnate the vegetable matter with water.

Danthanarayana has devised a method for the separation of the eggs of the weevil *Sitona* from the float by waterlogging the plant material by repeated freezing, filtering and transferring the whole float to a tube containing saturated sodium chloride solution; this is then centrifuged for five minutes (1500 r/m) and the eggs, and other animal matter, come to the surface whilst the plant material sinks. The eggs of some insects (e.g. the wireworm, *Ctenicera destructor*) sink rather than float, in sodium chloride (salt) solution because they are covered with small soil particles that have adhered to the coating from the female's glands. Doane found that this could be overcome by centrifuging in 20; alcohol solution, cleaning in 5 bleach (sodium hypochlorite) solution, followed by flotation and centrifuging in salt solution.

(3) *Grease film separation:* Using the same principle as above, namely the resistance of arthropod cuticle to wetting by water and its lipophilic nature, Aucamp devised a 'grease-film machine', in which greased glass slides are revolved in an aqueous suspension of the litter or soil sample: the arthropods adhere to the grease. However, small stones remove the grease, so impairing the efficiency of the greased surface and providing an alternative surface. Shaw devised an apparatus to overcome this problem, in which the soil sample is enclosed in a cylinder of coarse mesh in the centre of a plastic box; the lid of the box is greased with anhydrous lanolin and the box filled with water containing a little sodium hexametaphosphate. Several such boxes may be revolved on a wheel, after which the grease and arthropods, can be washed from the lid into a dish with hot soapy water or other solvent. Fragments and many whole arthropods may be recovered on a large scale from litter by Speight's 'greased-belt machine'. The litter is poured in aqueous suspension onto a moving (9 -12 in/ min) fine nylon bolting cloth belt (350 × 15 cm) (1.6 mm mesh), coated with petroleum jelly slightly thinned with liquid paraffin. At its lowest point the belt passes through a water trough when mineral

and plant debris falls off. Subsequently the belt, with the arthropod material now on its undersurface passes over a special rebated roller, under a powerful jet of water which washes the arthropods off into a tray. Speight found that extraction efficiency for whole arthropods or fragments was 70-90 %, except for whole organisms with diameters of less than 0.75 mm or greater than 10.0 mm for which it is unsatisfactory. About I litre of sample material can be extracted in one hour, the belt can run for 10 hours before 'regreasing' is necessary. The apparatus is particularly appropriate for dead organisms and fragments. To remove fine organic matter, Speight treated all samples with formalin, before wet sieving (1 mm mesh) and adding to the feeder tank.

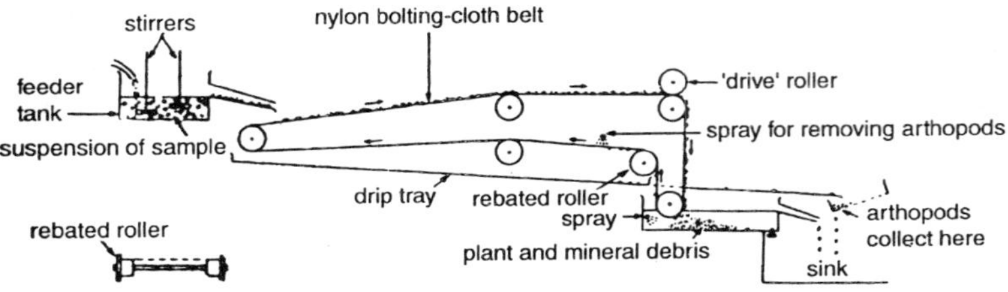

Figure 10.6 : *Speight's greased belt machine for the separation of arthropods from litter.*

Centrifugation

Although used widely in parasitological and pathological work, this technique has not been utilized to any extent in ecology, probably because only small samples can be treated at any time. However, it seems likely that it will be found to be of value for the final separation when the initial concentration has been by another method. Muller found that Acarina and Collembola could be more efficiently extracted from soil by centrifugal flotation in saturated salt (sodium chloride) solution than by certain funnels and, after initially separating mosquito eggs from soil by washing and flotation in magnesium sulphate, Service completed the extraction by this method. A similar technique has been used to separate nematode cysts from plant debris. Murphy gives a useful summary, in tabular form, of previous work using centrifugal flotation to separate animals from foodstuffs, soil and excreta.

Sedimentation

The principle of this technique is to utilize the difference in settling rates between animal matter and the substrate. It is the basis of an unpublished method of Davies for cleaning samples already separated from the soil; the sample is added to the top of a very long (4 m) glass tube, the base of which is in a rotating trough. However, the method is not very efficient as some mites settle at the same rate as soil particles. Seinhorst describes a sedimentation method ('Two Erlenmeyer method') for soil nematodes in which the flask containing the suspension was moved across a series of collecting vessels; the first of which was also inverted and the sediment in it further precipitated. The efficiency of the method was, however, only 60 - 75%.

Another sedimentation method for nematodes is described by Whitehead & Hemming, but their 'tray method' was generally more efficient.

Elutriation

This is an extension of the sedimentation process in which the sedimentation takes place against a water current flowing in the opposite direction and the principle is incorporated in two widely used methods for the extraction of soil nematodes: the Oostenbrink and Seinhorst elutriators. In the Oostenbrink model the nematodes are washed out of the sample and carried in suspension by the opposing water currents out of the overflow and through a series of sieves.

The Seinhorst elutriator is also part of a combined elutriation and sieving process; the soil is first passed through a coarse sieve and the suspension retained in a flask. This is inverted on top of the elutriator and the cork removed *in situ*; the upward flow of water (45 ml/min) coupled with the narrow sections A_2 and B_2 ensures that the nematodes are retained in sections A and B. After about half an hour the flask and the sections A and B may be drained through stopcocks A_3 and B_3. This material and that collected from the overflow is then wet sieved. A few large worms may pass into the soilcollecting container and this should be re-elutriated, but only for a brief period so that they will still be retained in vessel A. This method might well be adapted for single species studies on other organisms: those of a comparatively uniform size and mass, e.g. eggs, would be easily separated at a single level, whereas healthy, parasitized and dead individuals having different masses would separate at different levels.

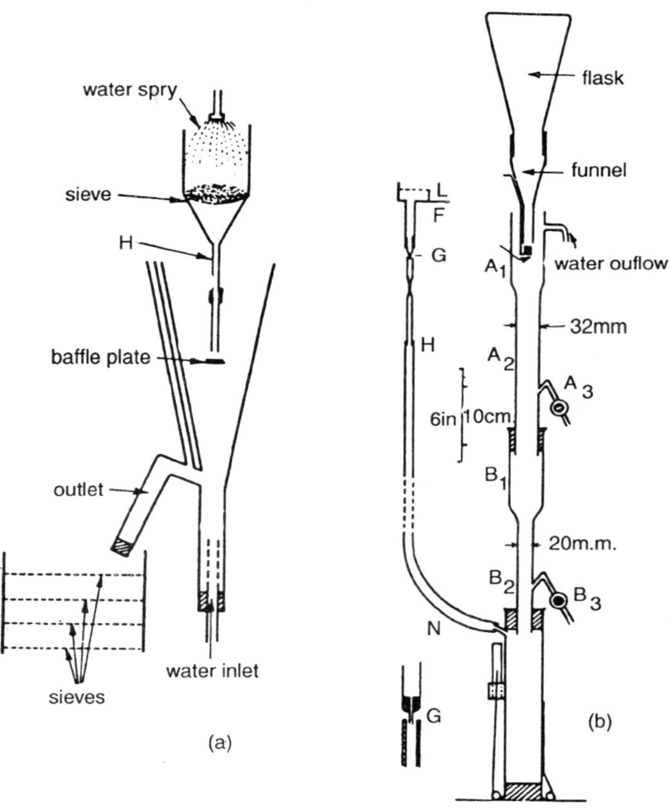

Figure 10.7 : *Elutriators:* a. *Oostenbrink's model;* b. *Seinhorst's model.*

Small arthropods such as Pauropoda and some Collembola that float in water may be extracted by von Torne's elutriator and sieving process in which they are carried to the top of apparatus.

Sectioning

Small animals may be examined and counted in soil sections.

Haarlev & WeisFogh have devised a method of fixing and freezing a soil core and then impregnating it with agar and sectioning. Gelatine is probably a more suitable substance for impregnation and an automatic method for this is described by Vannier & Vidal. Although the method is valuable for giving qualitative information on feeding sites and microdistribution, it is less suitable than other methods for population studies. Vannier describes how microscopic observations on the biology of soil animals may conveniently be made in an apparatus, incorporating frigistors (= 'frigatrons') that allow the sample to be rapidly cooled.

BEHAVIOURAL OR DYNAMIC METHODS

In these methods the animals are made to leave the substrate under some stimuli, e.g. heat, moisture (lack or excess) or a chemical. Their great advantage is that unlike the mechanical methods once the extraction has been set up it may usually be left, virtually unattended, and thus large quantities of material may be extracted simultaneously in batteries of extractors. Another important advantage is the ability to extract animals from substrates containing a large amount of vegetable material. The disadvantage is that, being based on the animal's behaviour, the extraction efficiency will vary with the condition of the animals and be influenced by changes in climate, water content, etc., experienced before and after sampling as well as by variations in these conditions in the apparatus itself. If samples have to be retained for several days, polythene bags seem the most suitable containers. Obviously eggs and other immobile stages cannot be extracted by this method.

The exact behavioural mechanism of extraction is not fully understood. Temperature gradients are established from the start, but in some types of funnel the humidity gradient is not clearly marked in the lower part of the sample. The pattern of egress of the mites is irregular and there is often a marked 'flush' towards the end of the extraction, this has been associated with the moisture content reaching about 20%, which triggers positive geotaxis. It seems that Oribatid mites are positively geotactic in dry conditions and negatively geotactic when the soil is moist;

the adaptive significance of such responses is obvious and they are probably an important part of the mechanism of the vertical movements of soil fauna. Whether all extractors act in this way, by the animal responding to a definite stimulus, rather than moving along a gradient, is uncertain. Many workers have sought to maximize the gradients and these extractors have been found efficient. There can be no doubt that for certain groups it is essential that the stimuli (usually heating and drying) are not applied too quickly, but with the diversity of soil organisms it is not surprising that it is impossible to construct an extractor that is equally efficient for all groups. Therefore although dry funnels, in particular, have been used extensively for community studies, their efficiency varies from soil type to soil type and according to the animal groups.

Dry Extractors

The basic apparatus is the Berlese-Tullgren funnel, a combination of the heated copper funnel designed at the turn of the century by the Italian entomologist, A. Berlese, and subsequently modified by the Swede, A. Tullgren, who used a light-bulb as a heat source. The funnel has been considerably modified and improved by many workers, as described in the reviews of Macfadyen and Murphy. Some of the most important innovations have been the demonstration by Hammer that when soil is being extracted the core should be retained intact and inverted, thereby enabling the animals to leave the sample by the natural passageways, and the discovery by Haarlov that serious losses could result from the animals becoming trapped in condensation from the core on the sides of the funnel; he recommended that the core should never touch the sides of the funnel and subsequent workers have sometimes referred to the space between the core and the sides of the funnel as the 'Haarlev passage'. When litter is being'extracted and the 'passage' is difficult to maintain, it may be helpful to increase air circulation with wide plastic piping. Large numbers of small funnels were first grouped together by Ford and this approach has been much developed and improved by Macfadyen, who also introduced the concepts of steepening the heat gradient and arranging a humidity gradient. Murphy, Newell, Dietrick, Schlinger & van den Bosch and Kempson, Lloyd & Ghelardi have all introduced devices to reduce the fall of soil into the sample, thereby ensuring a cleaner extraction. A folding Berlese funnel for use on expeditions has been developed by Saunders and a simpler model, in which the funnels are constructed of oil cloth and the animals 'forced down' by the vapour from naphthalene held in cheese cloth bags above the sample, has been devised by Brown.

There are many variants of the dry funnel in use, but most of them approximate to one of the following types.

Large Berlese Funnel

This is used for extracting large arthropods. e.g. Isopoda, Coleoptera, from bulky soil or litter samples and also for the extraction of insects from suction apparatus and other samples containing much vegetation. Desirable features are an air circulation system, introduced by Macfadyen, which may be opened to ensure a rapid drying for the extraction of desiccation-resistant animals, such as beetles and ants, or partly closed for a slower 'wet regime' (but avoid condensation) for beetle larvae, *Campodea* and other animals that are susceptible to desiccation. Macfadyen recommends testing the humidity below the sample by cobalt thiocyanate papers; for the resistant animals humidities down to 70% are tolerable, for the others it should not fall below 90%. As phototactic animals, e.g. Halticine beetles, are often attracted upwards towards the light-bulb of the funnel and so fail to be extracted, this is replaced as a heat source by a Nichrome or similar wire-grid heating element -this also heats the sample more uniformly. The extraction of positively phototactic animals has been further improved by Dietrick *et al.* who placed a 75 W spotlight below the collecting jar; this was switched on intermittently. An apparatus incorporating these and other features is illustrated in Figure 10.8. As Macfadyen and others have frequently stressed, for most animals the heat should be applied slowly and therefore itmay be useful to be able to control the voltage of the heating element through a rheostat or 'Simmerstat'. Extraction time will vary from a few days to over a month depending on the substrate and the animal.

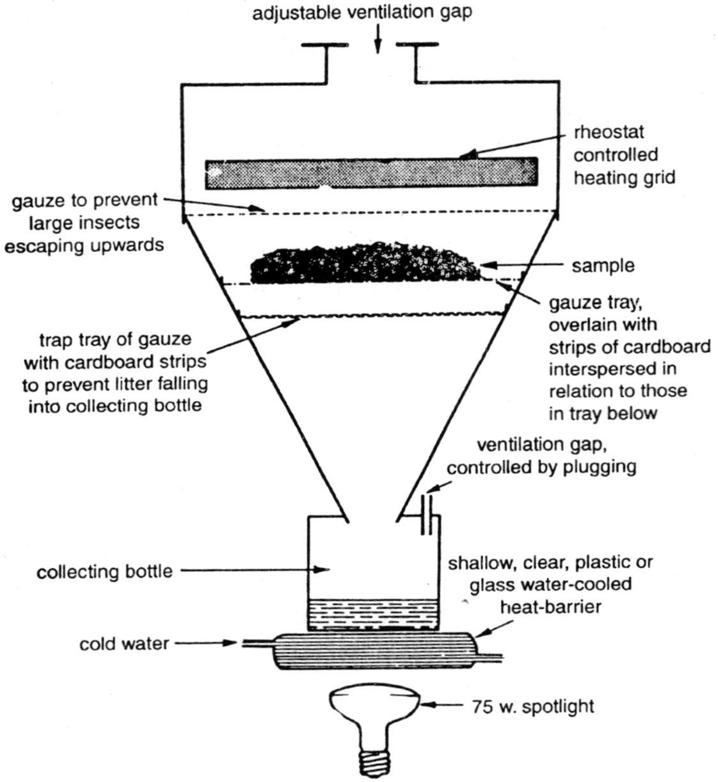

adjustable ventilation gap
rheostat controlled heating grid
gauze to prevent large insects escaping upwards
sample
gauze tray, overlain with strips of cardboard interspersed in relation to those in tray below
trap tray of gauze with cardboard strips to prevent litter falling into collecting bottle
ventilation gap, controlled by plugging
collecting bottle
shallow, clear, plastic or glass water-cooled heat-barrier
cold water
75 w. spotlight

Figure 10.8 : *A large Beriese funnel with modification.*

handle
hinge of lid of front extractor
adjustable shutter
samples
gauze covered ventilation holes
(a)
wire basket
(b)
collecting gutter
asbestos-cement block
500W. fire element

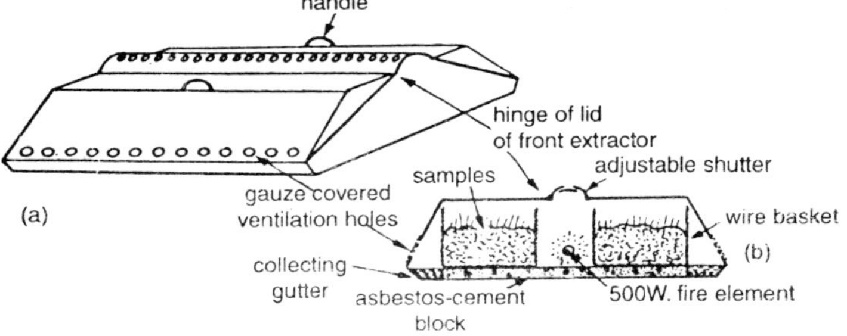

Figure 10.9 : *A pair of horizontal extractors : a. sketch; b. sectional view.*

Horizontal Extractor

Designed by Duffey for extracting spiders from grass samples, this extractor would probably serve equally well for any rapidly moving litter animal. It has the advantages that steep gradients are built up and that debris cannot fall into the collecting trough; it is also relatively compact and built in paired units. Heat is provided by a 500-W fire heating element controlled by a 'Simmerstat'. Air passes in through the ventilation holes just above the aqueous solution of the collecting gutters; this helps to prevent too rapid desiccation. Duffey recommended a 0.001% solution of phenylmercuric acetate, a fungicide and bactericide, with a few drops of a surface-active agent (detergent). The samples are held in wire trays 6 in (10 cm) wide. The extraction takes only a day or two.

High gradient (Multiple canister) extractor

Collembola, mites and other small animals may be obtained from small compact soil samples in this apparatus designed by Macfadyen. The samples are taken with the special corer, the samples removed, surrounded by the bakelite ring, and a sieve plate is fitted to each ring by pushing two springy wires into holes in the bakelite. The collecting canisters are cooled in a water bath

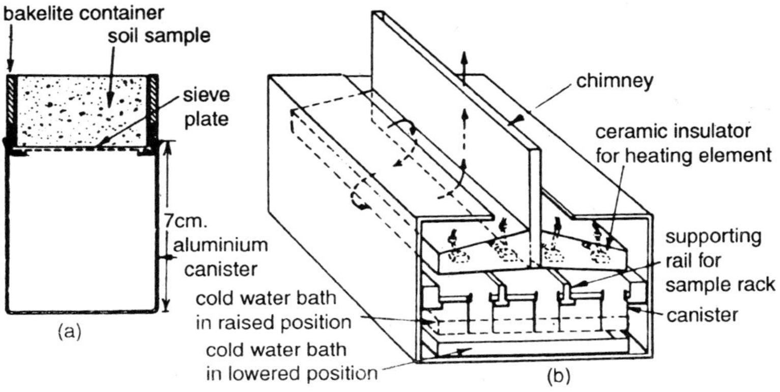

Figure 10.10 : *High gradient extractor: a. canister, core and sieve plate; b. whole apparatus.*

and 95 % humidity is maintained below the samples during extraction. Heat is provided by two 160-W elements giving 10 W per sample, and the temperature of the upper surface of the sample reaches 35°C; the temperature of the lower surface is determined by the water bath (e.g. 15°C), so that a very steep gradient is set up over the 3-cm deep sample. Extraction may be completed in 5 days, the maximum heat being applied after the first 24 hours. Macfadyen recommended a 0.3% solution of the fungicide 'Nipagin' (which contains streptomycin and terramycin) be used in the collecting canisters, but the tests of Kempson *et al.* indicate that a saturated aqueous solution of picric acid with a little photographic wetting agent might be preferable. Constructional details are given by Macfadyen, who has also described variants suitable for use under expedition conditions, the heating power being provided by kerosene or bottled gas, and the effect of various modifications on the physical conditions in the extractor. Block found that this apparatus extracted 76% of the acarina from mineral soils.

The Kempson bowl extractor

Developed by Kempson, Lloyd & Ghelardi, this method is suitable for the extraction of mites, Collembola, Isopoda and many other arthropods in woodland litter; extraction rates of 90-100% being recorded for groups other than the larvae of holom-etabolous insects for which it is not efficient. The apparatus consists of a box or shrouded chamber, containing an ordinary light-bulb and an infra-red lamp (250 W) that is switched on in pulses, at first only for aa few seconds at a time and gradually increased until it is on for about a third of the time. The pulsing of the lamp can be controlled by a 'Simmerstat', a time switch designed for electric. cookers, and the ordinary bulb is connected across it so as to protect the lamp from surges. The extraction bowls are sunk into floor of the chamber. Each bowl contains a preservative fluid and is immersed in a cold water bath. A saturated aqueous solution of picric acid, with some wetting agent, was used by Kempson *et al*, but Sunderland *et al.* have found a solution (at 80 g/1) of tri-sodium orthophosphate ($Na_3\ PO_4.12H_2O$) a cheap, safe and suitable alternative. The litter is placed in a plastic tray above the bowl and is supported by two pieces of cotton fillet net laid on a coarse plastic grid. The top of the tray is covered by a fine black

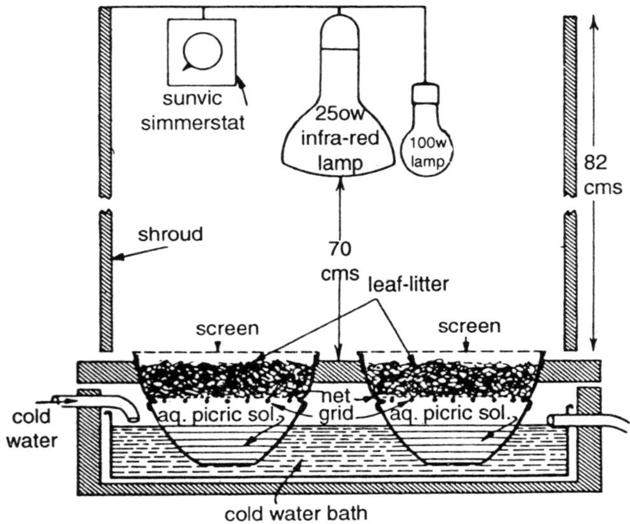

Figure 10.11 : *Kempson bowl extractor.*

nylon screen. Kempson *et al.* give very full instructions for the construction of this apparatus. Its great advantages, besides its comparatively simple construction, are the high humidity maintained on the lower surface of the sample, the lack of debris among the extracted material (but the fillet net does not appear to present any barrier to the animals), the prevention by the black screen of animals escaping upwards and the simple mechanism of gradually increasing the amount of heat to which the sample is subjected. Extractions with this apparatus generally take about a week.

Wet Extractors

The principle of this method is similar to that of the dry extractors, the animals being driven out of their natural substrate under the influence of a stimulus, possibly heat or, as the observations of Williams would suggest for some cases, reduced oxygen tension. As the substrate is flooded with water in all variants of the method there is no risk of desiccation and the method is particularly successful with those groups that are but poorly extracted by the Berlese-type funnels, that is nematode and enchytraeid worms, insect larvae and various aquatic groups. The method, which is faster than dry extraction, seems to have been discovered independently at least three times.

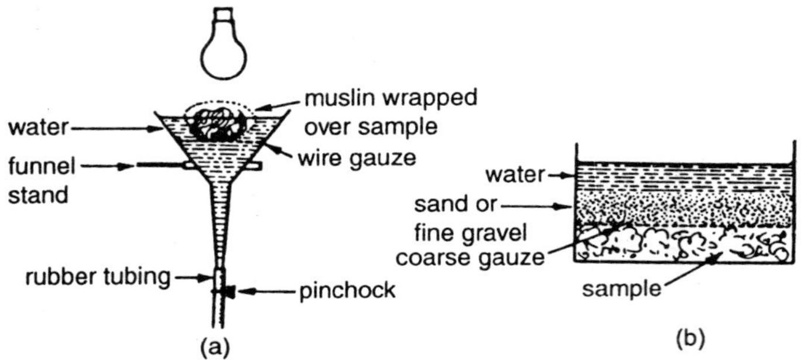

Figure 10.12 : a. ***Simple heated Baermann funnel.*** **b.** ***Sand extractor.***

Baermann Funnel

As originally designed this consisted of a glass funnel with a piece of metal gauze or screen resting in the funnel and a piece of rubber tubing with a pinchcock on the stem; the sample, which must not be very deep, is contained in a piece of muslin and partly flooded with warm water. Nematode worms leave the sample, and fall to the bottom of the funnel where they collect in the stem. They may be drawn off, in a little water, by opening the pinchcock. A number of refinements of this method have been introduced.

Nielsen ran a battery of Baermann funnels within a box with a lamp above; this heated the surface of the water in the funnels which was initially cold. Nematodes and rotifers could be extracted from soil and moss by this method, but it was not satisfactory for tardigrades. In order to extract Enchytraeidae, O'Connor used larger polythene funnels, spread the sample directly on to the gauze, without muslin, and submerged it completely; a powerful shrouded lamp heated the surface of the water to 45°C in three hours; dry samples must be moistened before being placed in the funnel and if the soil temperature is low samples must be gradually warmed to room temperature for a day. Extraction usually takes about three hours. In a series of tests O'Connor and Peachey have shown this method to be as efficient as the Nielsen inverted extractor (see below) for grassland soils and more efficient for peat and woodland soils. It is also very efficient for second and later instar tipulid larvae in moorland soils, but Kiritani & Matuzaki found only 10 % of the larvae of the cotton root knot eelworm, *Meloidogyne incognita* were extracted after a week.

Whitehead & Hemming introduced a modification that has been termed the 'Whitehead tray'. The soil sample is placed in a small (23 × 33 cm) shallow tray of 8 mesh/cm phosphor-bronze wire cloth, the inside of the tray being lined with paper tissue ('Scotties') and the outside supported by a plastic coated wire basket. This stands in a plastic photographic tray or similar vessel and is flooded with water until the surface reaches the underside of the soil (i.e. it is not inundated as much as in the Baermann funnel. After 24 hours at room temperature the wire basket is carefully lifted out and the water in the tray agitated and poured into vessels resembling separating funnels. After about 4 hours the nematodes collect at the tapered part of the funnel, which may be separated off by a rubber bung or diaphragm on a handle; the stop-cock is opened and the nematodes drawn off. The process may be repeated to further concentrate them. In comparative tests with clay, loam and sand and different nematodes, Whitehead & Hemming showed that this method usually extracted more efficiently than Seinhorst's 'Two Erlenmeyer method', elutriation or a sedimentation method they developed.

Hot Water Extractors

These have been developed by Nielsen for Enchyt-raeidae and by Milne, Coggins & Laughlin for tipulid larvae. Nielsen's apparatus involved the heating of the lower surface of the sample, contained in an earthenware vessel, by hot water; the animals moved up and entered a layer of sand, kept cool by a cold water coil, that was placed above the sample. The only advantage of this method over O'Connor's funnel is that it seems to extract more of the young worms; under conditions with a large amount of humus it is less efficient.

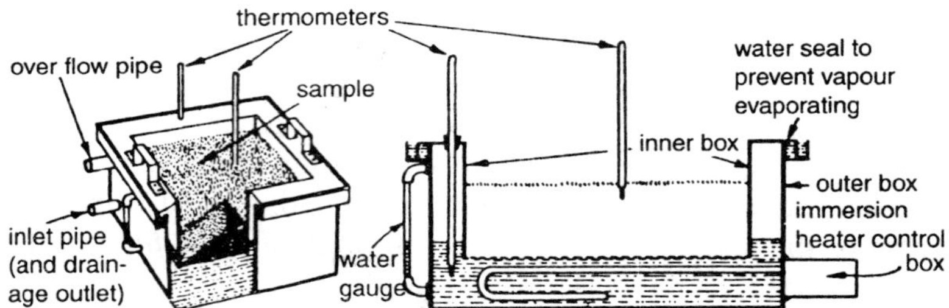

Figure 10.13 : ***Hot water extractor:***
a. ***sketch with segment 'removed';***
b. ***sectional diagram.***

The larvae of many insects, notably of some Coleoptera and Diptera, earthworms, molluscs, and most motile animals seem to be efficiently and rapidly extracted from soil samples if these are heated from below. SchjetzChristensen recorded 85% extraction of Elateridae using Nielsen's apparatus, modified by heating more strongly and omitting the cooling coil. Such an apparatus is not very different from that of Milne *et al.*, which was found to be virtually 100% efficient for active larvae and pupae. Essentially their device consits of two galvanized boxes, one within the other; the space between is a water bath heated by a thermostatically controlled immersion heater. The water temperature should not rise above 90°C. The inner box has a wire gauze base and holds the sample. The initial mode of operation is to fill the water bath until the level is just above the base of the sample, the heater is switched on until the temperature registered by the thermometer in the centre of the turf is 40°C; the water level is then raised to the level where it would just, eventually, flood the surface of the sample.

Heating and observation is continued, the insects being picked up as they leave the turf (but not before or they will retreat). The light above the extraction apparatus should not be too bright; Milne and co-workers recommended as a maximum, a 4-ft 40-W natural fluorescent tube 3 ft (91.5cm) above the sample; however, other species of insects might be more or less sensitive. Extraction times varied from an hour in a light soil to nearly three hours in heavy peat-covered clay. The chief disadvantage of this method as at present designed is that the apparatus must be watched continuously and the animals removed as they appear; however, equivalent mechanical methods usually take longer.

Sand Extractors

Extremely simple in design, consisting of a metal can, a piece of wire gauze and some sand, these devices seem to be remarkably efficient for extracting aquatic and semi-aquatic insect larvae and other animals from mud

and debris. The method was described by Bidlingmayer for the extraction of ceratopogonid midge larvae from salt marshes and has been improved and extended by Williams, who has found that the following freshwater invertebrates could be extracted from their soil substrate by this technique: Coelenterata, Turbellaria, Nematoda, Oligochaeta, Ostracoda, Acarina, Dipterous larvae, Gastropoda and Plecopoda. The sample, about 2 in thick, is cut so as to fill the container completely: it is then covered with 2 in of dry clean sand and the whole flooded with water. After 24 hours the sand may be scooped off and will contain large numbers of animals, but for 100 extraction of *Culicoides* larvae it should be left for 40 hours and for other groups tests would need to be made. The animals may be separated from the sand as Bidlingmayer and Williams recommend by stirring in a black photographic tray when the pale moving objects easily show up against the dark background. If the promise of this method is justified it will clearly repay further development. The final separation might be facilitated by the adoption of a sedimentation, elutriation or flotation method, aided perhaps by the replacement of the sand by fine washed gravel as used by Nielsen in his extractor. The chief disadvantages of the method in its present form is that it is difficult to make a clean separation of the sand from the soil after extraction; the placing of a piece of wide-mesh wire gauze between the two substrates might aid this operation, or if the extraction was done in a square vessel a plate could be pushed across between the layers at the end of extraction and the top layer poured off.

Cold Water Extractor

South (1964) found that slugs could be efficiently extracted from 1-ft (30.5 cm) deep turves by slowly immersing these in water. Initially they are stood in 1 in (2.5 cm) of water; after about 17 h the water level is raised to half the depth of the turf (i.e. 6 in, 15.2 cm); after 2 days the water level is again raised, this time in several stages, until at the end of another 24 hours it is within $\frac{1}{2}$ in (1.2 cm) of the top of the turf. This approach has also been used to extract the collembolan, *Hypogastrura,* and overwintering rice water weevils, *Lissorhoptrus*. The latter authors used a technique basically similar to the 'van Emden biscuit tin', but constructed from plastic bleach bottles, in which the sample is flooded. Such a technique is, of course, only suitable for robust insects.

Mistifier

This is essentially a method of extracting active nematodes from plant tissue. The plant material is spread on a tray, broken up and continually sprayed with very fine jets of water ('mists'). Combined with a cotton wool filter to collect the nematodes this method can be much more efficient than other techniques for certain eelworms.

Chemical Extraction

Chemical fumes may be used to drive animals from vegetation; the extension of such methods to soil and litter animals has been successful only with aphids and thrips. Chemicals in a liquid form, notably potassium permanganate solution, orthodichlorobenzene and formalin, often with a 'wetter', have been use for the extraction of insect larvae and earthworms. Formalin should be. applied four times, with 15 minute intervals, at the rate of 75 ml of 4 formaldehyde/13.5 l water/1 m^2 ground. Many workers have found these methods to be inefficient for earthworms; for tipulids in the grasslands studied by Milne *et al.* the efficiency was 85%. However the method can be efficient, depending on the type of soil, the species of animal and the season (which will influence the activity level of the animal). The more porous the soil and the more active the earthworm the more satisfactory the method. Among earthworms it is as efficient as hand-sorting for *Lumbricus terrestris* during active periods; on sandy soils it seems suitable for *Lirabellus* and *Dendrobaena octaedra*, but is seldom reliable for *Allobophora* spp.; where it is efficient it is more reliable for small worms than hand sorting.

Electrical Extraction

By discharging a current from a water-cooled electrode driven into the soil, Satchell was able to expel large numbers of earthworms; but, as he has pointed out, this method suffers from the disadvantage that the exact limits of the volume of soil treated are unknown and the efficiency of extraction seems directly related to the soil pH.

Index

O

P

R

S

T